Directions for Mathematics Research Experience for Undergraduates

Directions for Mathematics Research Experience for Undergraduates

Editors

Mark A Peterson
Mount Holyoke College, USA

Yanir A Rubinstein
University of Maryland, USA

NEW JERSEY • LONDON • SINGAPORE • BEIJING • SHANGHAI • HONG KONG • TAIPEI • CHENNAI • TOKYO

Published by

World Scientific Publishing Co. Pte. Ltd.
5 Toh Tuck Link, Singapore 596224
USA office: 27 Warren Street, Suite 401-402, Hackensack, NJ 07601
UK office: 57 Shelton Street, Covent Garden, London WC2H 9HE

British Library Cataloguing-in-Publication Data
A catalogue record for this book is available from the British Library.

DIRECTIONS FOR MATHEMATICS RESEARCH EXPERIENCE FOR UNDERGRADUATES

ISBN 978-981-4630-31-3

In-house Editors: V. Vishnu Mohan/Tan Rok Ting

Typeset by Stallion Press
Email: enquiries@stallionpress.com

Contents

Introduction

This volume stems from the conference called "New Directions for Mathematics Research Experiences for Undergraduates," held at Mount Holyoke College in June, 2013, celebrating 25 years of the Research Experience for Undergraduates (REU) program of the NSF. The majority of the articles reflect talks given at the conference. The conference itself is described by the organizers in the last article. The main impression left by the conference, documented and explored in these articles, is that the REU program in mathematics has stimulated innovations in mathematics education that go far beyond the original intent of the program. It is even possible to imagine that the sustained support of this program by the NSF over such a long period has begun to transform US mathematics itself.

The idea that undergraduates might do mathematical research as a part of their education has a long history, but in the early years it was a lonely position to take. A conference on this topic was held in the summer of 1961 at Carleton College, documented in *Undergraduate research in mathematics; report of a conference held at Carleton College, Northfield, Minnesota, June 19 to 23, 1961*, edited by May and Schuster [4]. A second conference on undergraduate research in mathematics, held in July of 1988, also at Carleton College, gave rise to a second volume, *Models for Undergraduate Research*, edited by

MSC 2010: 01A67, 01A80

Senechal [5]. In a section called "Voices from Long Ago" it reprints selections from the first volume.

In the years between these two conferences, several things happened that changed the terms of undergraduate research and its role in mathematics education. In the mid-1960s, as part of a continuing response to the Sputnik launch, the NSF created a program called Undergraduate Research Participation (URP), recalled by Lynn Steen in a contribution to an AMS publication *Proceedings of the Conference on Summer Undergraduate Mathematics Research Programs* (2000), edited by Gallian [2, pp. 331–336]. The aim of the URP was explicitly to accelerate the production of research scientists and mathematicians. Not long after, in the early Reagan years, there seemed to be a glut of research scientists, and the program was abruptly ended. By this time, though, the spark had caught. In 1977, Gallian initiated a summer program for undergraduate mathematics research that is still continuing today, and his example was widely noticed. In 1987, the NSF created the REU program.

The AMS volume cited above [2] contains articles from 35 REU and REU-like programs as of the year 2000, describing in detail how they operate. Increasingly, undergraduates have begun attending national meetings and giving talks, first in their own special sessions and conferences like Mathfest, but now also in general sessions. By 2006, Gallian could report "377 undergraduates attend annual joint meetings at San Antonio with 44 of them giving talks," as the culmination of a chronology in *Proceedings of the Conference on Promoting Undergraduate Research in Mathematics*, AMS (2007), edited by Gallian [3, pp. 267–272].

One thing that is not obvious in the above account is that the mathematics community as a whole was not especially eager to embrace these developments. Lynn Steen notes that in 1967, out of more than 500 URP grants, only 10 were in mathematics, and not because the NSF was discriminating against mathematics, but because most mathematicians were skeptical that undergraduate research made much sense. Every research mathematician works at the limits of his or her abilities, and these abilities have been hard-won over a lifetime. What would an undergraduate do? The

REU program — a program that spans all the sciences — aims "to attract a diversified pool of talented students into research careers" (NSF 96-102), but for many it was easier to imagine how this might happen in a laboratory science than in mathematics.

Arguably, if one sets the bar high enough, none of us does important research. In the bleak words of Alfred Adler, "Each generation has its few great mathematicians, and mathematics would not even notice the absence of the others. They are useful as teachers, and their research harms no one, but it is of no importance at all. A mathematician is great or he is nothing." [1]. Everyone understands what Adler is saying, but must also understand that this exalted notion of research is not the whole point, and that the interval between greatness and nothingness is far from empty.

The antecedents of the REU program were typically one brave professor and a few promising undergraduate students working on challenging problems. The early REUs continued this model for the most part, except that they were typically more like three brave professors and perhaps ten students working individually or in small teams. The undergraduates made interesting progress with reassuring consistency, and the basic idea began to take root. This history is considered in more depth, with observations about what it may mean for the culture of mathematics, in the first article of this volume, "Undergraduate Research and the Mathematics Profession," by Donal O'Shea.

The express objective of the REU program was to attract undergraduate students into research careers, but in practice the program produced other benefits as well, sometimes much more immediate. Directors of REU programs frequently noticed unanticipated dividends in their own research. This observation prompts the suggestion of the second article, "FURST — A symbiotic approach to research at primarily undergraduate institutions," by Tamás Forgács. It points out that many mathematics PhDs, trained in research, find themselves in teaching positions at small institutions that are essentially isolated from the research community. A slightly de-localized REU, organized around some suitable topic and including these small institutions, can reconnect these mathematicians to a community of like-minded professionals, with the undergraduates providing the "symbiotic" glue that holds the network together. This idea highlights the benefit not just to the undergraduates but also to their mentors.

Another benefit of the REU program, not expressly a goal but still an obvious consequence of it, is the enhanced learning that takes place in a research setting as opposed to a classroom. Modern research on learning and pedagogy makes clear the importance of active student involvement in learning generally, of all kinds, not just mathematics, and strategies for making learning more active are ubiquitous in innovative approaches to education in all subjects. Mathematics has a distinguished early contribution of this kind to point to in the "Moore method." It is thus no surprise that in the years of the REU program's existence there have also been strategies that put something like undergraduate mathematics research into the classroom. A good example of such a course, now beginning to be copied at other places, is MIT's, described in the third article of this volume, "A Laboratory Course in Mathematics," by Kathy Lin and Haynes Miller. The reader will notice much in common with the REU idea, even if the problems posed in the course are not precisely research problems. The importance of effective communication of mathematics both in speaking and in writing is heavily emphasized in this course, just as it is a big part of a typical REU.

The first experiments with undergraduate research in mathematics were typically at small undergraduate institutions, and this trend continues, by and large, in the REU program. On the one hand, it is not surprising that undergraduate colleges would be the ones most committed to an expressly undergraduate movement in education, but on the other hand it seems a little odd that research universities (R1s), the locus of most research in mathematics, would not be better represented. The fourth article in this volume, "REUs with limited faculty involvement, 'underrepresented' subjects in the undergraduate curriculum, and the culture of mathematics," by Yanir A. Rubinstein and Ravi Vakil, describes a program at Stanford that was, at least in part, a response to undergraduate demand. The new element, and the basis for the "limited faculty involvement," is graduate students. The reference to the "culture of mathematics" in the title is, in part, a consideration of the essential openness and generosity necessary for a research team of faculty member, graduate student, and undergraduates to function, possibly contrasting with the more private one-on-one interaction of a thesis advisor and a graduate student.

The role that graduate students might play in undergraduate research programs is further developed in the fifth article, "'The Berkeley Summer Research Program for Undergraduates': One model for an undergraduate summer research program at a doctorate-granting university," by Daniel Cristofaro-Gardiner. Both this article and the previous one point out that the areas of mathematics best represented in traditional REU programs are not typically the most active research areas at R1 universities. If the involvement of R1 universities and their graduate students in formulating problems and guiding undergraduate research is a growing trend, the mathematics of the REU programs may broaden to include subjects that are currently underrepresented there.

The sixth article, "Fifteen Years of the REU and DRP at the University of Chicago," by J. Peter May, points out another possibility for the REU at an R1 university: large scale. The presence of a thriving R1 research culture alongside the liberal arts college of the university has produced a remarkable fusion, in which graduate students and undergraduates work together on problems that sometimes constitute research, but always contribute to an active and energetic mathematical environment, if the statistics on the growth of the mathematics major is any indication. The DRP of the title is the Directed Reading Program, voluntary pairings of graduate students and undergraduates to read about some topic chosen out of pure interest, a format beginning to be copied at other places as well. The graduate students are compensated slightly, and the undergraduate students receive only the private personal benefit of the work that they put in, yet the program thrives. A DRP may sometimes metamorphose into a summer REU project, and the reverse may also happen. The success of the program, essentially restricted to

University of Chicago students for logistical reasons, may be judged by the size of the REU, about 80 students. REU students have considerable freedom to choose what they do, attending lecture courses that introduce research areas, choosing a problem, and meeting with graduate student mentors. Students are required to write up their work by the end of the summer.

The freedom of participants to choose their own problems is almost a defining feature of a long running program called the Mathematical and Theoretical Biology Institute (MTBI), founded at Cornell in 1996, now at Arizona State University, described in the seventh article, "Why REUs Matter," by Carlos Castillo-Garsow and Carlos Castillo-Chavez. One aim of this program, consistent with the long-term trend that continually finds new applications for the REU idea, is to recruit underrepresented minorities into mathematics. This means that it has to target students in their first or second year of study, at a level of mathematical experience considerably less than traditional REU students. The REU introduces, through lectures and exercises, the idea of modeling with differential and difference equations, but leaves it up to the participants to think of a problem that might be susceptible to modeling using these tools. The result is what the directors call a "reversal of hierarchy." Students choose situations for modeling that are more familiar to them than they are to their mentors, making them, in effect, the experts. Of course, the mentors will know more about the mathematical methods that the students will use. The results are impressive, with many mathematics PhDs resulting eventually over the years, undoubtedly traceable in part to this REU experience. The authors call attention also to the importance for future schoolteachers of experience using mathematics to investigate an applied problem of immediate interest.

The goal of attracting underrepresented minority students into mathematics is also one of the goals of the eighth article, "Integrating mathematics majors into the scientific life of the country," by William Vélez, but the strategy described there is ingeniously indirect. The author describes a mathematics major program designed to facilitate the participation of its students in the REUs of other sciences(!) It is pointed out that students of mathematics are frequently desirable participants in research programs outside mathematics, and that with a little forethought, such as course preparation in something like biology or geology, together with some computer experience, these students will move easily into applied research areas where their mathematical skills are immediately useful and valued. They may double major in mathematics and another science, or may even move into a research career in another science, leaving mathematics behind. From the point of view of the NSF and the REU program, this is still a success.

It is an historical fact that mathematics has lagged behind the other sciences in promoting undergraduate research for perhaps understandable reasons. Something of the sort may be taking place at present on a scale that, if the programs of the last several papers could be called "large," should now be called "vast." "The Gemstone Honors Program: Maximizing Learning through Team-based Interdisciplinary Research," by Frank J. Coale, Kristan Skendall, Leah Kreimer Tobin, and Vickie Hill, the ninth paper in this volume, describes a four-year program within the Honors College of the University of Maryland. Nurtured through a program that begins even before the opening day of their first term, teams of honors students learn to approach research, formulate their own problem, and bring it to a polished completion by the time they graduate. Building on years of experience, the program includes not only mentors expert in the subject matter of the research problems, but also specially designated librarians, and coaches for learning and practicing the skills of teamwork.

Mathematics, however, does not play much of a role in this structure, otherwise so seemingly comprehensive. One can think of reasons: undergraduates formulating an applied research problem that has special appeal for them will probably not think of mathematics playing much of a role. Once the problem is formulated in an unmathematical way, the absence of a role for mathematics becomes tautologically true. It is hard to imagine, though, that there should not be more place for mathematics, and that the problems and the education that students derive from this experience would not be thereby enriched. It would take initiative by mathematicians close to the scene, and also on the part of the directors, perhaps emulating the technique of the MTBI described above, introducing methods of mathematical modeling in an outline fashion before the research problems take their final shape.

An experiment of this kind has actually begun at the University of Texas at Austin in the program described in the tenth article of this volume, "The Freshman Research Initiative as a model for addressing shortages and disparities in STEM engagement," by Josh T. Beckham, James Farre, Gwendolyn M. Stovall, and Sarah L. Simmons. Like the Gemstone Program, the Freshman Research Initiative (FRI) is a very large scale program engaging undergraduates in research from the moment they arrive on campus. The FRI consists of around 25 "research streams" involving about 800 students for the first three semesters of their education. Despite the size of the program, spread across the sciences generally, there is little explicit involvement of mathematicians, and one cannot help but think that this is a missed opportunity for mathematics. The FRI, displaying remarkable agility, has recently taken steps to try integrating mathematics into its structure. The first experimental steps in this direction are described in this article, and the FRI itself is described to make it possible to imagine various approaches to integrating mathematics into such programs. The FRI and Gemstone are sure to be emulated elsewhere, and mathematics might play a different role if it were present at the creation.

The eleventh article is "Determining the Impact of REU Sites in the Mathematical Sciences" by Jennifer Slimowitz Pearl, writing from her experience as an NSF officer, and not as an official voice of the NSF. It is a remarkable fact that it is hard to know with any precision what the effect of the REU program in mathematics has been, despite its long and well-documented history, and despite the intriguing, informed speculations that are to be found throughout this volume. Students participate and fade away, often to re-appear as faculty colleagues. Directors of individual programs would naturally like to know how successful their programs have been, both in the short and the long term. The NSF even requires some such assessment each year, although the format is left open. The author proposes a flexible but still somewhat standardized tool for assessment (adapted from the geologists!) that might help bring clarity to the question of how effective the mathematics REU is.

The volume ends with a summary of the Mount Holyoke College conference.

References

[1] A.W. Adler, Reflections: Mathematics and Creativity, *New Yorker*, **47** (53) (1972) 39–45.

[2] J. Gallian (ed.), Proceedings of the Conference on Summer Undergraduate Mathematics Research Programs, American Mathematical Society, 2009, http://www.ams.org/programs/edu-support/undergrad-research/REU proceedings.pdf.

[3] J. Gallian (ed.), Proceedings of the Conference on Promoting Undergraduate Research in Mathematics, American Mathematical Society, 2007, http://www.ams.org/programs/edu-support/undergrad-research/PURMproceedings.pdf.

[4] K.O. May and S. Schuster (eds.), *Undergraduate research in mathematics: report of a conference held at Carleton College, Northfield, Minnesota, June 19 to 23, 1961.* Carleton Duplicating Service, Northfield, Minnesota, 1961.

[5] L. Senechal (ed.), *Models for Undergraduate Research in Mathematics*, MAA Notes Series No. 18, Mathematical Association of America, 1990.

Chapter 1

Undergraduate Research and the Mathematics Profession

Donal O'Shea
President's Office, New College of Florida
5800 Bayshore Road, Sarasota, FL 34243, USA
doshea@ncf.edu

1. Introduction

Much has been said and written about undergraduate research in mathematics: about its possibility, about its intrinsic quality (or lack of), about different models for facilitating it, and about its value to students, mentors, and even departments. Very little has been written about the systemic effect of undergraduate research on the mathematics profession and the development of mathematics as a whole. What follows is a first, necessarily speculative, attempt to begin to fill this gap.

2. Historical Sketch

I begin by remarking that mathematics as a discipline, and as a profession, in the US is comparatively young. Although higher education dates back to before the American Revolution, it was difficult to get a good graduate education in mathematics in 1900 in the US (and virtually impossible in 1890). A student with a serious interest in mathematics had little choice but to go to Europe,

---

MSC 2010: 01A61, 01A80 (Primary); 01A65, 01A67 (Secondary)

especially Germany or Paris. This would change in the beginning of the twentieth century as some of the educational, social, and technological ferment that followed the Civil War took root. The nation needed professionals and postsecondary education grew rapidly. The Morrill Act established the land grant institutions and in the three decades from 1869 to 1899, the number of institutions of higher education increased from 563 to 977, enrollments quintupled to just under a quarter of a million students and bachelor degrees awarded tripled to over 27,000.[1] The new universities needed professors. Some older, more established, American institutions, notably Harvard and Yale, responded by establishing programs to train university professors with disciplinary expertise. Johns Hopkins University, Clark University, and the University of Chicago were set up to emulate the best German practices. The emergence of the American mathematical research community parallels the emergence of American research university out of the German research model. This is a fascinating story, well documented in [22].

The new mathematics research community soon began to experiment with better methods of graduate education. R.L. Moore began teaching topology in 1911 to beginning graduate students using a method that basically mimicked research, and others followed suit. Efforts to encourage undergraduate research followed shortly after. In 1934–35, the Michigan Section of the MAA appointed a committee to create a wider interest in undergraduate research, and subsequently opened space for undergraduates to present at their annual meetings (see [26]). In [10], Frank Griffin surveyed college and university mathematics departments to find out which were involving undergraduates in research, where he defined "research" as "exploring some new question or re-exploring some old question and getting results previously unfamiliar to specialists in the field." He documented an impressive array of student research achievements, including those at his home institution, Reed College, and noted that colleges that encouraged

[1]The number of doctoral degrees granted in all subjects went from 1 in 1869 to 362 in 1899 (see, [17, Table 301.20]).

their undergraduates to do research produced more students who went on to receive graduate degrees in mathematics.

The point I make here is that undergraduate research in mathematics, although not common in the early decades of the twentieth century, is nearly as old as mathematical research in the US. A few individuals had noted its efficacy, observing that students who had been involved in research as undergraduates were more likely to continue in mathematics and were, in turn, more likely to involve undergraduates in their research. Some sectional MAA student meetings encouraged student presentation. Some mathematics departments required student theses. Others, however, had a comprehensive exam, a practice almost orthogonal to student research.

The late 1950s to mid-1970s and the 1960s in particular, mark the greatest period of change in higher education in the United States since the post-civil war period, save perhaps the present. The exuberance of the times coupled with the large increase in college enrollments as a result of the GI bill and the post-World War II baby boom fueled huge growth and diversification in American higher education. Enrollments in postsecondary institutions more than doubled in the 1960s, from 3.6 million enrolled in 1959 to more than 8 million 1969, and tripled from 1960 to 1975, reaching 11.2 million in 1975. See [17, Table 313.10]. The number of institutions grew more slowly, but still dramatically, increasing 25% in the 1960s (from 2004 in 1959–60 to 2556 in 1969–70) and an additional 10% in the first half of the 1970s (reaching 2765 in 1975–75). See [17, Table 317.10]. The Sputnik launch by the Soviet Union in 1957 and the subsequent press, not to mention the 1960 presidential campaign, produced a sense that the US needed to catch up by producing more scientists, engineers and mathematicians. Federal research dollars flowed into the system and the American university morphed into what Clark Kerr, first chancellor of UC-Berkeley and president of the University of California system, called the multiversity: not one community, but a community of communities [14].

From the point of view of mathematics and undergraduate research, this era was more notable for what did not happen than what did. Federal research agencies, especially the then recently established (1950) National Science Foundation funded a number of experiments to complement mathematical and science education. Among them, was the URP (undergraduate research program), which funded undergraduate research through the 60s and 70s. This funding, however, was primarily focused on biology, chemistry and physics, not mathematics, and allowed students to spend time, typically ten weeks, in active research laboratories, at research universities and some liberal arts colleges. The program did not impose the requirement that departments accept students from other institutions. Chemists at liberal arts colleges started incorporating more summer research into undergraduate programs.

In mathematics, the projects funded by the National Science Foundation tended to involve enrichment activities aimed at complementing formal mathematical education. Most involved accelerated instruction, coupled with a format that encouraged students to work together and competitively on challenging problems. Perhaps the best known such program is the Ross Summer Mathematics program for high school students which originally began at Notre Dame in 1957 and which moved to Ohio State University in 1964 where it has been ever since. A number of government labs provided research opportunities in which mathematics students could participate, but there seems to have been little pressure from the mathematical community to create summer research opportunities similar to those in the other sciences. There were a number of projects to institute academic year programs and to improve the major. In particular, a program directed by Kenneth May at Carleton College was explicitly aimed at producing undergraduate research. The project involved selecting particularly strong entering undergraduates and supplying them with a rich set of problems and a social environment to encourage research in their junior and senior years. It set up an undergraduate colloquium joint with St. Olaf [15].

The first conference on undergraduate research in mathematics of which I am aware took place at Carleton College in 1961. The Summary and Resolutions of the conference committee made it clear that the prevailing assumption was that ordinarily undergraduates could not be expected to do significant research in mathematics (see [27]).

"[The aims of undergraduate research] are the training and stimulation of the student, not the attainment of new results, though such bonuses will come occasionally." The document goes on to state that "Undergraduate research should be judged by standards different from those now employed by mathematicians" and notes that using the word "research" may in fact inhibit the establishment of undergraduate research programs at colleges and universities, suggesting that one might broaden the appeal by calling undergraduate research "independent study." Nonetheless, the resolutions of the conference carried the clear sense that "student activity of the research type was a part of good educational practice" and were aimed at the best means for achieving it.

So, there were disputes about what precisely "undergraduate mathematical research" meant and others who questioned its value altogether, asserting that it came at the expense of time that would be better devoted to"mastery of established topics." Nonetheless, some faculty members in mathematics departments, both in liberal arts colleges and in research universities, were encouraging undergraduates to do research, usually but not exclusively as part of a senior thesis. (See [10], [11] and [26], as well as the articles [28], [23] and [16], reprinted in [25].) There was considerable anecdotal evidence that undergraduate research enhanced the likelihood that students would continue to graduate studies in mathematics. It was clear, however, that even in a time of relative plenty, there was neither sufficient appetite nor consensus to use research dollars to fund undergraduate research. In retrospect, this seems to have been a missed opportunity for the mathematics community.[2]

In 1977, Joseph Gallian established a summer research program in mathematics at the University of Minnesota Duluth. See [6]. That program, which is still running today, brought a small number of undergraduates from across the country for ten weeks to work on

---

[2]During this same time, of course, the funding pattern characteristic to mathematics of concentrating research support on relatively small individual investigator grants became entrenched. Over time, this would result in the mathematical sciences being critically underfunded in comparison with the other sciences, a situation addressed in the so-called David Report of the early 1980s. See [18], and its successor [19].

unsolved problems, with a goal of obtaining and publishing new results. In other sciences, of course, it had already been accepted that undergraduates could contribute usefully to the research enterprise.

The 1980 presidential election brought a new ethos to the country as the growth of preceding decades tapered off. Perceptions of an oversupply of scientists abounded and the URP program was terminated in 1981. By the mid-1980s, alarm had arisen at the decrease in the number of young people going into science and mathematics careers. The David Report [18] documented a critical shortage of young people entering mathematical careers. Data gathered by the American Chemical Society showed a 25% decrease in the number of B.S. level chemists graduated during the early 1980s. Congress gave the National Science Foundation a mandate to create a plan to counter the trend.

One result was the Research Experiences for Undergraduates (or REU) program which began to solicit proposals in 1987 for the so-called summer research sites. Sites were encouraged to tailor programs to suit their own calendars, and to make efforts to increase populations of women, Native Americans, minorities and other under-represented groups (see [24]). Mathematicians with regular NSF research grants were also allowed to tack on REU supplements that would allow funding one or two undergraduates as part of an investigator's research program. This program resulted in the establishment of a number of REU sites in mathematics, one organized by an alumnus of the Duluth program. Some of these sites still exist — in fact, it is the twenty-fifth anniversary of the REU site at Mount Holyoke College (see [21]) that has occasioned the present conference.

In July 1988, a conference on undergraduate research in mathematics was organized at Carleton College in connection with the Second National Conference of the Council on Undergraduate Research.[3] The proceedings, entitled *Models for Undergraduate Research in Mathematics* and edited by Lester Senechal, surveyed some older, as well as some new, summer programs. Although the issue of whether

[3]The Council on Undergraduate Research is an organization that had been established in 1979, largely by chemists at liberal arts colleges, to promote undergraduate research.

undergraduates could do research had not entirely been laid aside, a number of the new sites focused on having undergraduates working together to create new mathematics. Some of the sites exploited the fact that advances in algorithms and computing speeds had made it possible for undergraduates to explore examples that were at the forefront of what was known theoretically.

This time around, enough mathematicians embraced undergraduate research to ensure that it would flourish. A mathematical and computer science division was added to the Council on Undergraduate Research in 1989. In addition, research institutes such as MSRI that had been established by the National Science Foundation as an attempt to enhance the infrastructure for the mathematical sciences began to incorporate undergraduate research. Shortly thereafter, the National Science Foundation started funding sites and departments to vertically integrate mathematical education and research. Although NSF subsequently discontinued the formal VIGRE program, undergraduate research played an important role in these projects. Undergraduate research sites also became an important component of a number of programs to enhance the participation and retention of women and under-represented minorities in mathematics.

Today, there can be no question that undergraduate research is firmly established. Many colleges and universities have found ways to institutionalize programs, either with gifts or by convincing university administrators to allocate funds to support undergraduate research. A recent article by Joseph Gallian [9] details a number of impressive data points. At the time of writing, the NSF is funding 59 REU sites. Indeed, one could argue that undergraduate research is one of the features characterizing American mathematical education.

3. Undergraduate Research and the Profession

As we have just seen, undergraduate research in mathematics is now well-established and has been taking place over the last twenty-five years. What can we say about its effects on the profession? Oddly enough, despite the ubiquity of undergraduate research, there has been curiously little effort to assess its cumulative effect on the profession. On the one hand, this is not surprising. Site directors are busy, and they care about mathematics, and their students — so what assessment there is, happens at the site level. There have been relatively few attempts to aggregate the data from different sites, or to survey mathematicians.

Here then is an attempt to examine some of the effects that undergraduate research has had on the profession, together with a rueful acknowledgement/warning that the absence of careful studies renders much of what I say speculative.

The time since 1985 has seen steady growth in higher education. Postsecondary enrollments have increased from 12.2M in 1985 to 20.6M in 2012, and the number of institutions from 3340 to 4726. The number of bachelor degrees awarded over the same period has increased by 81%. Somewhat disturbingly, the number of bachelor degrees in mathematics and statistics has not kept up. There were 16,122 bachelors with majors in mathematics and statistics in 1985 and 18,842 in 2011, an increase of only 17%. This is often "blamed" on computer science, but bachelors in computer and information science increased even less: there were 42,337 degrees awarded in 1985–86 compared to 47,384 in 2011–12, an increase of 12%, even less than mathematics and statistics. This is in sharp contrast to baccalaureates granted in business, communications/journalism, health, parks/leisure, which increased 55%, 100%, 150% and 743% respectively to 366,815, 83,777, 163,440 and 38,993, respectively. These numbers, of course, hint at some of the profound structural changes occurring in higher education in our times.

Despite the anemic increase in the number of mathematics and statistics baccalaureates, the number of doctoral degrees in mathematics increased 71% from 978 in 1990 to 1669 in 2012. We have just seen that this increase cannot be attributed to enrollment growth. In fact, undergraduate degrees in mathematics and statistics actually declined slightly from 1985–86 to 2006–07. Although one would have to investigate more carefully, it is not unreasonable to conjecture that the increase is due to undergraduate research programs.

Undergraduate research interacts with higher education as a whole, and with the mathematics profession, in other ways as well. At the time of writing (2014), there are about 4800 institutions of higher education in the United States, which together enroll about 21 million students, about 90% of whom are undergraduates (see [17, Tables 313.10, 317.10; IPEDS database]). A large plurality of students attends primarily associate degree granting institutions. There are more than 1900 of these (over 600 of which are for profit) enrolling 8.2 million students. Best known are the institutions with significant doctoral programs. These enroll 5.8 million students, and there are nearly 300 of them, enrolling on average 20,000 students. The next largest category is the 700 master's degree granting institutions which enroll around 5 million students. The smallest sectors are the primarily baccalaureate degree-granting institutions, which enroll 1.4 million students in slightly more than 800 institutions (averaging 1757 students each), and the 805 specialized and tribal institutions which enroll 0.8 million students. (See [3].)

A substantial portion of students who go on to receive a doctoral degree in mathematics receive their baccalaureate degrees from two much smaller subcategories of schools: the hundred or so schools comprising the so-called National Research One Universities (R1's) and the hundred or so National Liberal Arts Colleges. These terms harken back to the older Carnegie classification that was in place from the early 1970s to 2004, and widely used (and misused) by various constituencies with interests in higher education, among them popular publications such as US News and World Report. The Research 1's have significant overlap with what Carnegie now lists as the 108 "very high" research universities, which enroll on average about 26,000 students per year. The National Liberal Arts category, perpetuated by US News and World Report, includes another 100 colleges, made up of residential primarily baccalaureate institutions that grant over 80% of their degrees in the arts and sciences, together with some purely undergraduate, residential engineering schools that require students to complete a liberal arts core (such as the national military academies, Harvey Mudd and Rose-Hulman).

The National Research 1 Universities and the National Liberal Arts Colleges together award less than a third of the bachelor degrees granted in the United States. Yet over 70% of those who received doctorates in science and engineering fields received baccalaureates from these institutions. (See [12], and the later updates [2,5].) Among these institutions, the top 25 research universities and top 25 liberal arts colleges vastly out-produced the others.

In mathematics, the patterns are both more striking and more complicated. Over half of those who received doctoral degrees in 1990–95 received their undergraduate degrees from foreign institutions (see [12, p. 3]). The half who received baccalaureate degrees in US institutions were concentrated among even fewer institutions than the other sciences. The top 25% of research universities and top 25% of liberal arts colleges produced more than 40% of baccalaureates receiving PhDs in mathematics from 1990–95. In contrast to engineering and the other sciences with the exception of chemistry, liberal arts colleges contributed proportionately more baccalaureates who received mathematics doctorates than the research universities. In fact, one college, St. Olaf, produced more baccalaureates who received PhDs in mathematics (29) from 1991–95 than any research university save UC Berkeley, Harvard and MIT (with 62, 50, and 42 baccalaureates receiving mathematics PhDs from 1991–95 respectively). Reed College, with 24 baccalaureates in the period 1991–95, out-produced all but six universities (the three just listed, and UCSD, UCLA, and the University of Wisconsin-Madison, with 28, 27, and 24, respectively). See [12, Appendices]. Women who subsequently received PhDs in mathematics were even more likely to have received baccalaureates from liberal arts colleges, and in particular from the handful of women's colleges (Bryn Mawr, Mount Holyoke, Smith and Wellesley). By contrast, blacks who received doctorates in mathematics were much less likely to have received bachelor degrees from HBCUs than in engineering and other sciences. (See [12, Table 17].)

Needless to say, these figures raise the troubling possibility that whether a student decides to pursue a career in mathematical sciences may have more to do with the accident of where he or she receives her undergraduate education than with his or her interest or ability. Although we again lack a systematic study, there is evidence to support the claim that undergraduate research sites have begun to play a role in democratizing good mathematics education. The NSF currently sponsors 59 sites. Two thirds are at institutions that are neither Research 1 universities nor liberal arts colleges. Such institutions educate large numbers of students, but have played proportionally less of a role in sending students to graduate school in mathematics.

We now have a generation of mathematicians nearing their forties, a substantial proportion of whom have been through undergraduate research programs. I think that some of the changes in the culture and conduct of mathematics can be traced to that generation and to undergraduate research in particular. For example, thirteen of the 59 REU sites funded in 2014 are at Research 1 universities. Twenty-five years ago, directing an REU site at a research university would have been considered professionally risky. That clearly has changed. Indeed, this conference makes it clear that several additional undergraduate research efforts at research universities are being supported by REU supplements and department funds. The individuals directing these latter efforts had benefited from undergraduate research experiences earlier in their careers. It is plausible to conjecture that another systemic effect of undergraduate research programs has been the creation of a cadre of faculty/researchers willing to supervise undergraduate research.

Mathematics has always been social, although mathematicians have not always described it as such. Today mathematics is more deliberatively collaborative and mathematics departments and tenure committees do not blink at joint papers. Many of our leading mathematicians have, or participate in, blogs and many of them had been in REU programs as undergraduates. Although, it would be absurd to credit undergraduate research programs solely with the increasingly collaborative nature of mathematical research today, I have little doubt that they have played a role.

I also believe that REU sites and other forums in which undergraduate research occurs have improved the quality of undergraduate mathematics education at many institutions, including those that already graduate a substantial number of mathematics and statistics majors who persist to receive doctorates. Advisors at many Research I schools and liberal arts colleges actively encourage interested students to participate in REU sites. Indeed, some liberal arts colleges have refashioned their mathematics majors knowing that undergraduates will pick up certain skills by participating in research experiences. (See [20].) Undergraduate research has also had some effects on curriculum. The need to provide references for undergraduate research students has given rise to a number of attractive books.

To conclude with one more speculative effect of undergraduate research, let me note that skeptics often say that undergraduate research is a myth, at best relegated to those areas of mathematics which do not require much knowledge to contribute (and that even then the contributions are marginal). Fields mentioned include graph theory, combinatorics, finite group theory. "Mainstream" areas in analysis, number theory, algebraic geometry and topology, by contrast, the argument goes, are necessarily inaccessible to those without substantial graduate training. Consider, however, the areas of mathematics that have seen the most massive development in the last two decades. These include combinatorics, graph theory, and computational commutative algebra and are areas in which undergraduates have been most involved in research in the last twenty-five years. Might the increased involvement of young people through undergraduate research be partly responsible for the rate of progress? There can be no question that some of the rapid development stems from the increasing impact of computing in those areas of mathematics, and some of the slower development in other areas from the high barrier to entry of other fields. Nonetheless, the coincidence is worth noting.

4. Conclusion

We are currently witnessing a rapid, global evolution in the mathematics profession. The society that will do best is the one that will attract and develop the most talent. We have also seen that undergraduate research is a distinctive piece of the mathematics enterprise in the United States. In addition to the benefits it confers on students, which every program director can recite, it has changed the profession and practice of mathematics in this country. Conjecturally, it has played a role in increasing entry into the profession despite flat undergraduate enrollments, in democratizing the profession, in encouraging women and under-represented minorities to pursue mathematics, in helping redefine the mathematics major, in encouraging the publication of accessible texts in under-explored areas of mathematics, in making the practice of mathematics more collaborative, in supplying a new generation of active researchers interested in teaching and mentoring undergraduates, and in directly and indirectly enhancing progress in certain areas of mathematics. See [9] for some data.

A little over a decade ago, Ani Adhikari and Deborah Nolan [1] pointed out the need to carefully assess undergraduate research programs. The same need exists to document the effect of undergraduate research at a more macro and systemic level, and it is not difficult to imagine how to construct studies that would help assess the conjectural claims made above. In an era in which policy is increasingly framed in terms of maximizing short-term easily measured outcomes, funds directed to activities with longer term, less predictable outputs are increasingly at risk. The conjectural account outlined above suggests that curtailing undergraduate research would inflict profound damage over the next two decades on the mathematical infrastructure in the US. We would do well to remember that federally funded undergraduate research disappeared in 1980 almost overnight. It is in all our interests to do the studies that might help prevent a similar decision two or more years hence.

References

[1] A, Adhikari, D. Nolan, "But What Good Came of It at Last" How to Assess the Value of Undergraduate Research, *Notices Amer. Math. Soc.*, **49** (2002) 1252–1257.

[2] J. Burrelli, A. Rapoport, R. Lehming, Baccalaureate Origins of S&E Doctorate Recipients, InfoBrief, Science Resources Statistics, Directorate for Social, Behavioral, and Economic Sciences, July 2008, Washington: National Science Foundation, NSF 08-311.

[3] Carnegie Foundation for the Advancement of Teaching. Summary Tables. 2013. http://classifications.carnegiefoundation.org/summary/ugrad_prog. php.

[4] F. Connolly, J.A. Gallian, What Students Say About Their REU Experience, in *Proceedings of the Conference on Promoting Undergraduate Research in Mathematics*, Providence: American Mathematical Sociaty, 2007, pp. 233–236.

[5] M.K. Fiegener, S.L. Proudfoot, Baccalaureate Origins of U.S.-trained S&E Doctorate Recipients, InfoBrief, National Center for Science and Engineering Statistics. Washington, DC: National Science Foundation, NSF 13-323, 2013, http//www.nsf.gov/statistics.

[6] J.A. Gallian, The Duluth Undergraduate Research Program, in *Models for Undegraduate Research in Mathematics,* Mathematical Association of America, 1991, pp. 15–18.

[7] J.A. Gallian, *Proceedings of the Conference on Summer Undergraduate Research Programs.* Providence, RI: American Mathematical Society, 2000, http://www.ams.org/employment/REUproceedings.pdf.

[8] J.A. Gallian, *Proceedings of the Conference on Promoting Undergraduate Research in Mathematics.* Providence, RI: American Mathematical Society, 2007.

[9] J.A. Gallian, Undergraduate Research in Mathematics has Come of Age, *Notices Amer. Math. Soc.*, **59** (2012) 1112–1114.

[10] F.L. Griffin, Undergraduate Mathematical Research, *Amer. Math. Monthly,* **49** (1942) 379–385.

[11] F.L. Griffin, Further Experience with Undergraduate Research, *Amer. Math. Monthly,* **58** (1951) 322–325.

[12] S.T. Hill, Undergraduate Origins of Recent (1991–95) Science and Engineering Doctorate Recipients, Special Report, 1996, Division of Science Resource Studies, Directorate for Social, Behavioral and Economic Sciences, Washington: National Science Foundation NSF 96–334.

[13] F. B. Jones, The Moore Method, *Amer. Math. Monthly,* **84** (1977) 273–277.

[14] C. Kerr, *The Uses of the University*, Cambridge: Harvard University Press, 1963, Fifth Edition, 2001.

[15] K.O. May, Undergraduate Research in Mathematics, *Amer. Math. Monthly,* **65** (1958) 241–246.

[16] K.O. May, Heretical Thoughts on Undergraduate Research, in *Carleton 1961 Conference*, reprinted in *Models for Undegraduate Research in Mathematics,* Mathematical Association of America, 1991, pp. 187–191.

[17] Digest of Educational Statistics, National Center for Education Statistics, 2013, http://nces.ed.gov/programs/d13/tables.

[18] National Research Council, *Renewing U.S. Mathematics: A Critical Resource for the Future.* Washington, DC: The National Academies Press, 1984. (The "David Report.")

[19] National Research Council, *Renewing U.S. Mathematics: A Plan for the 1990s.* Washington, DC: The National Academies Press, 1990.

[20] D. B. O'Shea, H.A. Pollatsek, Are Prerequisites Necessary? *Notices Amer. Math. Soc.*, (May 1997), 210–221.

[21] D.B. O'Shea, Summer Research at Mount Holyoke College, in *Models for Undegraduate Research in Mathematics,* Mathematical Association of America, 1991, pp. 31–37.

[22] K.H. Parshall, D.E. Rowe, *The Emergence of the American Mathematical Research Community, 1876–1900: J. J. Sylvester, Felix Klein, and E. H. Moore.* Providence, RI: American Mathematical Society, reprinted with corrections 1997.

[23] P.C. Rosenbloom, An Undergraduate Research Seminar, in *Carleton 1961 Conference*, reprinted in *Models for Undegraduate Research in Mathematics,* Mathematical Association of America, 1991, pp. 175–191.

[24] K.B. Schowen, Research as a Critical Component of the Undergraduate Educational Experience, in *Assessing the Value of Research in the Chemical Sciences: Report of a Workshop*, (Chapter 7). Washington, DC: National Academies Press, 1998.

[25] L.J. Senechal, *Models for Undergraduate Research in Mathematics.* MAA Notes, No. 18, Mathematical Association of America, 1991.

[26] E.R. Sleight, Undergraduate Research in Michigan, *Amer. Math. Monthly*, **48** (1941) 696–697.

[27] S. Schuster, K.O. May, Summary and Resolutions, in *Carleton 1961 Conference*, reprinted in *Models for Undegraduate Research in Mathematics,* Mathematical Association of America, 1991, pp. 157–159.

[28] R.L. Wilder, Material and Method, in *Carleton 1961 Conference*, reprinted in *Models for Undegraduate Research in Mathematics,* Mathematical Association of America, 1991, pp. 161–174

Chapter 2

FURST — A Symbiotic Approach to Research at Primarily Undergraduate Institutions

Tamás Forgács

California State University, Fresno, CA 93740, USA

The FURST program is designed to address two commonly cited problem areas of a traditional REU: (i) great variations in students' level of preparation, and (ii) the difficulty of following up with the completion of the research after the summer program is over. The program serves faculty as well as student interest, in order to address concerns about the benefits of involving junior faculty in undergraduate research mentoring.

1. Introduction

Over a quarter century ago, the first REU programs in mathematics started involving undergraduate students in research. Since then, the number of active REU sites has grown to about fifty, engaging somewhere around 400 students in research in any given summer. The growth in the number of participating students over the last three decades was followed by the expansion of the concept of meaningful research conducted by undergraduates. Today, there are dozens of ways in which undergraduates are engaged in research, including partnerships with companies and high schools, working on problems ranging from the highly theoretical to the contemporary applied.

MSC 2010: 01A65 (Primary); 01A67 (Secondary)

At the core of a rewarding research experience lies a well-selected research problem. Such problems have two important characteristics:

(i) the research student has sufficient mathematics background so that she/he can start working on the problem with a few hours of focused and specialized additional instruction, and
(ii) the student has a high probability of making significant progress on the problem within the given time frame.

Thus, two obvious constraints on accessibility of a research problem are the length of the REU program, and the academic preparation of the participating students.

The original approach of many REUs was to choose problems for their importance and interest to the research community. There is great value in this approach: faculty interest and enthusiasm for the project are all but guaranteed, and students are partaking in the discovery and development of the frontiers of the discipline. Arguably, this is the best way to get students excited about mathematics, and to steer them towards a career as a research mathematician. Research active faculty mentoring students with exceptionally good mathematics preparation can easily overcome both of the above-mentioned constraints, and can benefit greatly from selecting and designing REU projects this way.

With the growth in the scope and interpretation of the undergraduate research experience, REU sites today provide access to research opportunities for a wide variety of students, including those whose mathematics background is at the sophomore/early junior level.[1] Such programs may find great value in student-oriented (and student involved) problem selection.

If the first approach aims to involve students in a faculty-driven endeavor, we can view the second as one involving faculty in a student-driven endeavor. We may see the importance of the latter

[1]W. Velez collected information on REU programs which admit students who have not yet taken upper division mathematics courses, and also on programs which otherwise suit specific student needs. For the complete list (last updated in January, 2014) see [16].

grow, especially if the undergraduate research community's efforts of institutionalizing undergraduate research (UR) prove fruitful in the coming years. In order to achieve the goal of making UR a permanent fixture on the landscape of curricula, we must pay attention to the interests of both major constituents involved in the enterprise, especially at primarily undergraduate institutions (PUIs).

The rest of the paper is organized as follows. In Section 2, we discuss junior faculty at PUIs, and their goals pertaining to supervising undergraduate research. Section 3 outlines the benefits and shortcomings of a traditional REU, and Section 4 describes a possible answer to some of those shortcomings. We conclude with Section 5, discussing the ongoing pilot project, as well as the prospect of future funding, and the future of the project as a whole.

2. Junior Faculty and Their Career Goals

One of the perhaps lesser known statistics about mathematics research is the median number of research papers authored by mathematicians: two.[2] This is an often shocking number to recent mathematics PhDs as they are used to being immersed in the vibrant research environment of an R1 university, where two papers (or more) could be the annual production of each active research faculty. While there are several reasons one can think of as for why this number is what it is, we mention here just one: the majority of mathematics PhDs find jobs somewhere other than a research university. In particular, many are hired by PUIs,[3] which educate roughly 60% of all undergraduate mathematics students.[4] Depending on the particular institution, junior faculty at PUIs may or may not

[2] This was the number as of 2005 (see [5]) and is unlikely to have increased since given the growing number of mathematicians coupled with shrinking number of available positions in academia.

[3] As described in [11], in 2010 over half of the advertised positions requiring math PhDs were at PUIs in the sample (Groups M and B), and this fraction was closer to 65% in 2007.

[4] To be more precise, 59% of undergraduate mathematics degrees awarded in 2010–2011 were granted by Groups M and B, with group B (institutions granting only baccalaureate degrees) showing the largest increase in the number of awarded degrees [2].

be required to conduct research in addition to teaching and service duties.

Consider, as a snapshot of the current research expectations at PUIs, the California State University system,[5] comprised of 23 campuses strewn across the state. In a recent survey, we asked the chairs of the mathematics departments to answer the following two questions:

(i) Is conducting research a requirement for tenure in your department?
(ii) What quantifiable ways does your campus/department assess whether a tenure track faculty has met the research requirements for tenure?

The data we gathered indicate that research and creative activity is a requirement for tenure at all campuses that responded (16/23). Of these, ten would consider the publication of two research papers in peer-reviewed journals as satisfying the research requirement (recall the statistic mentioned earlier!), four require three articles, and two ask for six or more peer-reviewed papers during the probationary period.[6] Encouraging signs of institutions recognizing faculty involvement in undergraduate research are present at two CSU mathematics departments. One department considers the establishment of an REU site and mentoring in the REU the equivalent of a peer-reviewed research paper for tenure purposes. The other has clearly established ways to count student research supervision towards eligibility for research assigned time. Still, we need to do much better in crediting faculty involved in UR for the work they do, a change the REU community has been promoting for some time.

Unfortunately, conducting mathematics research is not quite like riding a bicycle: once one stops, re-engaging takes more time, and

[5]By no means do we claim that this is a representative sample of PUI institutions. The CSU is a very large system with all the positive externalities that come with size. There is, however, a decent variation in the location of campuses (urban/rural), and in the size of the individual institutions and their mathematics (or mathematics & statistics) departments.

[6]One of the institutions probably should not be considered a PUI, as it grants PhDs in several of the sciences, although not in mathematics.

hence becomes less and less likely as time goes by. If the mathematics community does not want the research talents of hundreds of mathematics PhDs to shrivel and die on the diverse academic grapevine (and hence severely limit the return on the considerable investment of a 4–6-year degree), it might be worthwhile to think about what we can do to keep the junior faculty at PUIs engaged in research in a sustainable way.

Some believe that early involvement with undergraduate research can be detrimental to a junior faculty's career. Such an opinion is perhaps grounded in the current evaluation systems at PUIs, as well as in the assessment that undergraduate research at PUIs involving students with less than exceptional preparation is not likely to produce co-authored research publishable in peer-reviewed journals. Anecdotal evidence also suggests that junior faculty may have a difficult time establishing an undergraduate research culture at a PUI if it does not already have one. Possible difficulties include, but are not limited to, obtaining funding for a new REU site, and finding the time to carry on with one's research program while supervising undergraduate students in research. The latter is especially difficult if the student research projects are in areas and at levels vastly different from those of the faculty. We posit that junior faculty comprise a resource for the UR endeavor which we should exploit as much as possible. Is there a way to serve junior faculty and enhance their career goals while involving them in undergraduate research? We would like to think so.

3. The Heritage of the REU Program

Some of the impressive outcomes of the REU program at large include students writing research papers, giving talks and poster presentations, PhD programs recognizing the value of students having had such an experience, and employers looking for new hires who express their desire to get involved in ongoing student research programs and initiatives. By demonstrating the validity and value of undergraduate research, REUs greatly contributed to the expansion of undergraduate research activities from the summer to the academic year. The

Center for Undergraduate Research in Mathematics (CURM, [15]) at Brigham Young University is a fantastic example of such an expansion. The program is in its second five-year cycle, with the primary purpose of supporting students engaged in research during the academic year. In addition to CURM, there are several other ongoing efforts of integrating UR into the curriculum (see Miller [Ch. 3], Beckham [Ch. 10] and Cole [Ch. 9] for example). These are great programs, but they do not have the advancement of faculty research as a primary objective.

The longevity of the REU program has allowed directors and faculty mentors to evaluate many aspects of the undertaking, and identify areas where improvements could be made. The varying levels of student preparation and the restricted length of the average REU program are two such areas.

Providing access to research experiences for a student body with increasingly diverse backgrounds has many great advantages. It does however raise the problem of assembling research groups with relatively closely matched talents and mathematics preparation. REU mentors can hardly afford the luxury of delivering individually tailored instruction (especially if the number of students in a group exceeds 3 or 4) during the summer program. Faculty might be more inclined to make such an investment of time and effort in student development, if the allotted time for both training and conducting research were longer. Extended length programs would also be in a better position to complete more comprehensive research projects. Continuing with the mentoring of students after they leave an REU site is very difficult, and typically the writeup of the results is the activity that suffers most from this constraint. Affording longer time to complete research projects may in turn increase the number of research papers produced, and with that the visibility of undergraduate research results to the mathematics community.[7]

[7] One must absolutely pay tribute to the organizers of the undergraduate poster sessions at the JMM and MathFest, along with the organizers of all forums dedicated to the exposure of undergraduate research. We believe that mathematicians reading research

As much as we have increased access to REUs, recent conversations among the directors clearly indicate a severe shortage in spots currently available to students. While we all wish we could have more mathematics REU sites,[8] the economic reality is that increasing the number of mathematics REU sites significantly would require a substantive additional allocation to the REU pot of the Division of Mathematical Sciences within the NSF, an ill-fitting measure with the current federal funding trend. Thus it seems unlikely that REUs in their current format could meet the rising demand, and they certainly have continuing difficulty addressing the above-mentioned challenge areas. On the other hand, established REUs have figured out how to run an effective program, which workshops for students are most useful, how to bring in outside speakers, and in general how to provide the logistics necessary for a well-rounded research experience. There is tremendous value in this accumulated know-how.

We believe that given the right framework, we can address some of the shortcomings of the current REU framework. Moreover, we can do so by allowing undergraduate research students and their mentors scattered throughout the country to enjoy the benefits of an established REU site without duplicating efforts or having to run an REU themselves.

4. FURST — A Possible Answer

The *Faculty and Undergraduate Research Student Teams* (**FURST**) program aims to involve faculty and students at primarily undergraduate institutions in collaborative mathematics research in a way that

papers while unaware of the authors being undergraduate students legitimizes UR in the strongest of ways, and should be one of the goals the enterprise continues to aspire to achieve.

[8]The NSF's REU sites webpage currently lists 69 sites in chemistry and 144 sites in biology [12, 13]. Although the number of chemistry and biology REU sites is larger than those in mathematics, when considering the percentage of students majoring in these areas, math is not being disproportionately underrepresented with its 49 current sites [1].

addresses some of the shortcomings of a traditional REU. In particular, the program provides opportunities for

(1) collaborative research among junior faculty at PUIs so as to help them achieve pertinent career goals in research;
(2) a year-long, faculty-mentored undergraduate research experience for students at PUIs; and
(3) a think-tank like summer immersion experience for each member of the FURST team at an existing REU site.

By involving existing REUs, the program profits from the positive externalities provided by these sites in the form of workshops, speakers, and other activities which make for a well-rounded research experience for the participating students. While serving as the location for the month-long immersion phase, the REU sites also provide FURST faculty with the infrastructure to support an intensive face-to-face research effort with their peers. We believe that such an opportunity for faculty can effectively produce research results in a timely fashion, as evidenced by the success of programs like the AIM SQuaREs and the IAS Park City Mathematics Institute, among others. In addition, by utilizing existing REU infrastructure, we address the problem of faculty at smaller PUIs not being able to attract external funding for a research program due to the size of their department and institution.

Finally, by extending the scope of the research project to a calendar year, we provide students and faculty with an opportunity not only to engage and complete more difficult problems, but also to introduce and promote undergraduate mathematics research at the institutional level at their respective colleges and universities. Although not a primary goal of FURST, the extended research period may positively impact the number of students obtaining advanced degrees in the mathematical sciences:

> [in STEM] 30% of the researchers with more than 12 months of research experience reported that they expected to obtain a PhD, compared to only 13% of those with 1 to 3 months of research experience and 8% of those with no research experience. (See [9]).

4.1. Program implementation

All logistical and organization aspects of the program are handled at California State University, Fresno, including the publication of a list of research areas (along with their REU host sites) for the ensuing three-year period, the administration of the application process, and the selection of FURST teams.[9] From the applicants' point of view, a typical cycle of the program starts with a faculty member applying to one of the research areas identified for the upcoming calendar year. Along with a short CV, each applicant has to provide FURST with a few broadly described research problems, some proposed for faculty collaboration, some for student research.

Support for participants is comparable to the REU levels of support. We give students one month stipend for the immersion period, and support their travel to the REU host site, and to the Joint Mathematics Meetings. Faculty receive a summer stipend, while the home institution of each FURST team receives funds towards a course buyout. Applicants are required to submit a letter of commitment from their chair or dean guaranteeing the course release in exchange for the funds. We regard this as a small but important step towards institutionalizing undergraduate research, by making administrators aware of such possibilities and their associated costs (Figure 1).

Once selections have been made, faculty members recruit two students at their institutions to become members of their team. There are two aspects of faculty selecting students from their own institution which we consider important: (i) faculty tend to know students at their schools more in depth, and are able to evaluate them more completely, when compared to evaluating a student based solely on

[9]All relevant information about the program will be published in [14], which, at the writing of this article, lists the details of the pilot year.

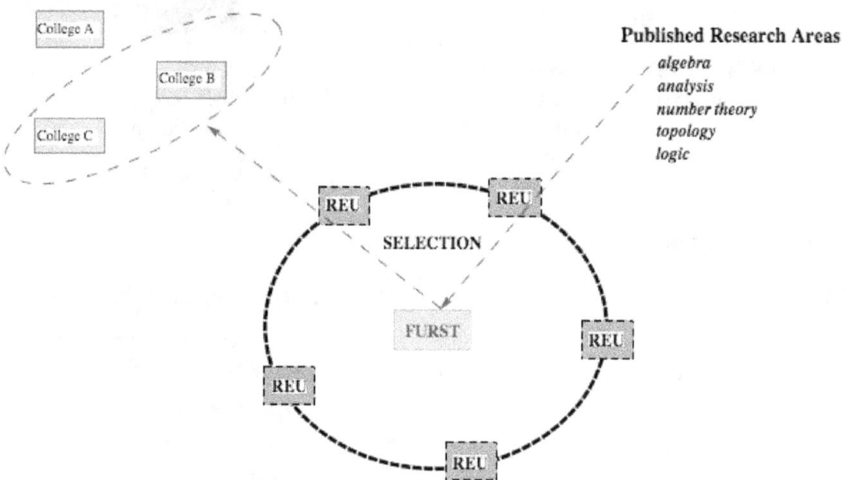

Figure 1. An example of a possible list of research areas, and selection process. Faculty from Colleges A, B and C apply to FURST in the research area of algebra.

their application packet and letters of reference; (ii) students who would hesitate to commit to work with faculty they do not know at a place they are unfamiliar with may be more amenable to work in a foreign environment with a faculty they know well.[10]

During the spring semester of the calendar year, FURST faculty begin the research collaboration with their counterparts, as well as the mentoring of their own students. The goal of the spring mentoring is to make sure that by the end of May, students' preparation is adequate for their research project. In order to help with certain aspects of mentoring (especially for junior faculty), the program has arranged for its faculty to participate in a three-day workshop organized by CURM at the end of May. This workshop provides, along with many other resources for faculty and their students, some ideas about effectively working with undergraduate research groups [15]. Teams begin research in earnest during June, emulating a typical REU schedule.

[10]There is extensive literature on the psychology of leaving home, or a familiar environment in order to attend college (see for example [3, 4, 7, 8, 10]). On a smaller scale, one can view the circumstances of going to a summer research experience similar to those of going away for college. We believe that familiarity with the faculty may help students overcome some of the arising issues.

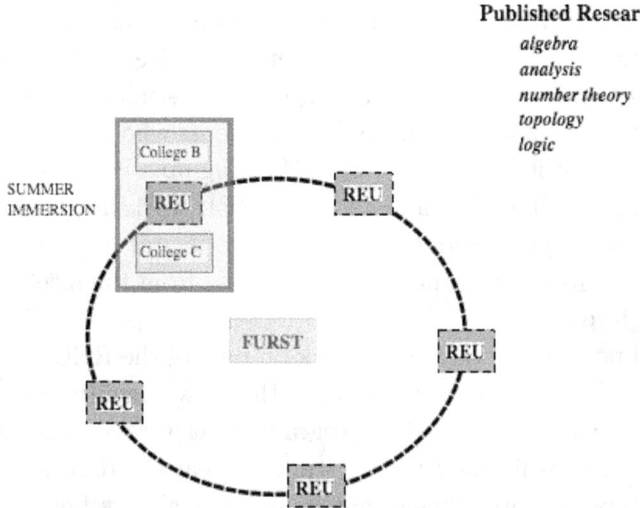

Figure 2. Example continued: FURST teams from Colleges B and C are on site with the host REU for one month during the summer.

In July, which is usually the second month of an REU, all FURST teams meet at an existing REU site for a one-month long immersion experience (Figure 2).

The REU sites hosting FURST cohorts will have been chosen years in advance so that REU and FURST research topics would be in concert. The primary reason for coordinating research topics is to ensure that FURST students could be integrated in the developing REU scientific community with ease. During the immersion, FURST teams reap the benefits of the REU tangibles: student presentations, outside speakers, professional development workshops and communicating with peer researchers. It is during this immersion phase that FURST faculty have a month of intensive face-to-face research time with their collaborators. The one-month length of the immersion phase may also address the issue of some students not wanting to be away from their homes and families for an entire summer, and therefore not participating in research experiences at all.

We recognize the possibility of the FURST teams' arrival at the host REU site being perhaps disruptive in terms of the developing dynamics of the REU program. Certainly, if the REU students have formed groups and have reached a social balance by week four (which they usually do by then), the new students may feel like outsiders. On the other hand, the arrival of a new group of students may serve as a chance to start anew for those REU students who have not integrated as well as others for some reason. Setting up a Facebook page, and including the FURST students in the REU workshops from the beginning (via Skype or some other similar software) may eliminate some, or all, of the awkwardness that may arise from the insertion of a new group into the REU mix. From a logistics perspective, FURST teams would not require any extra work on part of the REU mentors or the PI. FURST students come with their own faculty mentors, and they only participate in REU programs that were already set up to run. Finally, housing the FURST students near the REU students would be a plus, but we expect that it may not always be possible, which is certainly a challenge (Table 1).

At the end of the immersion phase, FURST teams return to their home institutions. During the fall semester, faculty continue to work with their students, who are expected to complete their research

Table 1. Program implementation timeline.

Dates	Activity	Location
August–January	Application process	N/A
First 2–3 weeks of February	Selection process	N/A
Spring semester	Assembly of FURST teams, student selection, problem selection both for students and faculty collaboration, exposition to background material	Home institution
June	Research experience of the FURST teams begins	Home institution
July	Immersion experience	Host REU site
Fall semester	Continuing research as a “dedicated course”, completion of the written report, preparation of talks and posters	Home institution
January of following year	Presentations and reunion	Joint Math Meetings

projects by the end of the calendar year. To facilitate the follow through, the program provides funds for, and requires that FURST faculty receive a one course buyout in the fall semester. The typical cycle ends with students submitting their work for publication, FURST teams presenting their work at the Joint Mathematics Meetings, and faculty submitting a research paper for publication.

An expressed goal of the program is that primarily undergraduate institutions would recognize the merit, and support the development of research active faculty. As the Council on Undergraduate Research notes: "... [an] important reason for faculty members at PUIs to maintain an active and productive research program is because of the positive educational outcomes for the undergraduate student collaborators ...". Despite its importance, "... the significant barriers ... at many PUIs may limit the ability of faculty members at these institutions to conduct transformative research ..." [6]. FURST hopes to address these issues while exploiting the wonderful resources established REU sites have to offer.

5. The Present and the Future

FURST obtained funding to run a pilot program in 2014. The inaugural teams are from Washington College and Fresno State, with the host REU site also at Fresno State. The two faculty in the pilot program participated in the CURM faculty workshop at the end of May of 2014. They both reported on tremendously positive experiences there, giving a preliminary justification for keeping this workshop as part of the FURST program. We have also concluded the immersion experience of the inaugural FURST cohort. The integration of the FURST students into the REU program went very smoothly both from an academic, and from a social point of view. We did feel however that bringing the FURST students for the first four weeks of the REU program might make it easier for them to integrate socially, and will entertain this option for the future years. We put together a proposal for funding FURST for a three-year period starting 2015. We hope that the proposal is successful, most of all so that we would be able to provide research opportunities to junior faculty and their students at primarily undergraduate institutions, which they may not otherwise have. The pilot year of FURST is supported by the National Science Foundation grant # DMS-1156273.

References

[1] X. Chen, STEM Attrition: College Students' Paths Into and Out of STEM Fields, NCES 2014-001, National Center for Education Statistics, Institute of Education Sciences, U.S. Department of Education, Washington, DC.

[2] R. Cleary, J. Maxwell, C. Rose, "Fall 2011 Departmental Profile Report" in the Annual Survey of the Mathematical Sciences in the U.S., http://www.ams.org/profession/data/annual-survey/2011Survey-DepartmentalProfile-Report.pdf (Accessed on 5/13/2014).

[3] J.M. Dennis, J.S. Phinney, L.I. Chuateco, The Role of Motivation, Parental Support, and Peer Support in the Academic Success of Ethnic Minority First-Generation College Students, *J. College Student Development*, **46** (3) (2005) 223–236.

[4] A.R. Desai, The Psychosocial Experience of First Generation Students Leaving Home for College, *Professional Psychology Doctoral Projects*, Paper 20, 2012.

[5] J. Grossman, Patterns of Research in Mathematics, *Notices Amer. Math. Soc.*, **52** (1) (2005) 35–41.

[6] K. Karukstis, N. Hensel, (eds.) *Transformative Research at Predominantly Undergraduate Institutions.* Council on Undergraduate Research, 2010.

[7] H.B. London, "Breaking Away: A study of First Generation College Students and Their Families, *Amer. J. Education*, **97** (2) (1989) 144–170.

[8] J.S. Phinney, K. Haas, The Process of Coping Among Ethnic Minority First Generation College Freshmen: A Narrative Approach, *J. Social Psychol.*, **143** (6) (2003) 707–726.

[9] S.H. Russell, M.P. Hancock, J. McCullough, Benefits of Undergraduate Research Experiences, *Science*, **316** (2007) 548–549.

[10] M.C. Zell, Achieving a College Education: The Psychological Experiences of Latina/o Community College Students, *J. Hispanic Higher Education*, **9** (2) (2010) 167–186.

[11] "Survey Results (Faculty Recruitment)", http://www.ams.org/profession/employment-services/Fac-recruit-surv.pdf (Accessed on 5/13/2014).

[12] NSF REU sites: Chemistry, http://www.nsf.gov/crssprgm/reu/list_result.jsp?unitid=5048 (Accessed on 5/13/2014).

[13] NSF REU sites: Biological Sciences, http://www.nsf.gov/crssprgm/reu/list_result.jsp?unitid=5047 (Accessed on 5/13/2014).

[14] Faculty and Undergraduate Research Student Teams program site, http://www.fresnostate.edu/csm/math/research/furst/index.html (Accessed on 5/12/2014).

[15] Center for Undergraduate Research in Mathematics, http://curm.byu.edu/ (Accessed on 5/9/2014).

[16] Undergraduate Research Opportunities, http://sites.williams.edu/Morgan/2012/01/21/undergraduate-research-opportunities/ (Accessed on 5/13/2014).

Chapter 3

A Laboratory Course in Mathematics

Kathy Lin
Graduate School of Education, Rutgers University
New Brunswick, NJ 08901, USA
ywklin@gmail.com

Haynes Miller
Department of Mathematics
Massachusetts Institute of Technology
Cambridge, MA 02139, USA
hrm@math.mit.edu

For the past eleven years, the Mathematics Department at the Massachusetts Institute of Technology has offered an undergraduate course known as 18.821 Project Laboratory in Mathematics. The course is designed to give students a sense of what it is like to do mathematical research. It exhibits characteristics of undergraduate research programs and Inquiry Based Learning courses, but it does not fit comfortably in either camp. This paper describes the course and its reception by faculty and students, and explores its relationship to contemporary trends in learning theory.

1. Introduction

One of the degree requirements for undergraduates at MIT is completion of one of the twenty-four courses satisfying the Laboratory Requirement, described as follows:

> The Laboratory Requirement is not intended primarily to teach specific techniques for later experimental work, provide broad coverage

---

MSC 2010: 97D50, 97D40 (Primary)

> of a particular field, or complement a specific subject. The laboratory subjects are planned to give each student . . . an opportunity to work on one or more experimental problems, exercising the same type of initiative and resourcefulness as a professional would in similar circumstances.

Up until 2004, MIT Mathematics majors had to satisfy this requirement outside the department. The most common choice was 15.310 Managerial Psychology Laboratory, offered by the MIT Sloan School of Management. For some of our majors, this was an appropriate choice; for most, it was not. In the spring term of 2004, the Mathematics Department launched its first — and thus far only — approved lab subject, 18.821 Project Laboratory in Mathematics. Our idea was to base the course on the scientific method as it is practiced in mathematics. For this, we had to devise a way to give students something of the experience of doing mathematical research, but in a sustainable manner, repeatable term after term with a wide variety of students. See [5] for an account of the creation of this course.

The course strategy was devised by Mike Artin and the second author, following ideas of Michael Brenner (now Professor of Applied Mathematics and Applied Physics at Harvard University). It has not changed in any essential way since its inception. Students work in teams of three on three open-ended projects, under the mentorship of experienced researchers. They explore puzzling and complex mathematical situations, generate and examine data, define research directions, search for regularities, and attempt to explain them mathematically. Students share their results through professional-style papers and presentations.

The expectations of the course are quite different from those of most others that students take. There is no set content for the students to master. Not only are there no right answers; there aren't even any right questions! The structure of the Project Lab is distinct from that of a true Research Experience for Undergraduates, and the goals are different as well. We might summarize the course goals as follows. As the course progresses, each student will

- develop research skills, including exploring examples, choosing and narrowing goals, and communicating findings;

- experience mathematics from the bottom up, starting with observing patterns and seeking to discover mathematical explanations for them;
- realize that mathematics is a living, evolving field to which the student can contribute;
- enhance mathematical understanding by putting earlier learning to use in new settings;
- improve written and oral communication skills, including use of language and argument in formal and informal settings;
- consolidate teamwork skills such as dividing tasks, being responsive, and taking responsibility for the work of the team.

While the Project Lab has obvious connections with Inquiry-Based Learning, it is separated from that movement by the absence of target disciplinary skills or understandings; the central course objectives are purely experiential. It is said that in the "Moore method" nothing is given — the student has to discover everything — but in a sense everything is given; the student is supposed to come to a predetermined conclusion. The Project Lab has an opposite philosophy.

Beyond the MIT context, the Project Lab adheres to a key recommendation of the forthcoming 2015 Curriculum Guide to Majors in the Mathematical Sciences. In this guide, the MAA Committee on the Undergraduate Program recommends that:

> Every major student should work, independently or in a small group, on a substantial mathematical project that involves techniques and concepts beyond the typical content of a single course. Students should present their results in written and oral forms. Institutions can provide this opportunity in various ways, or a combination thereof: undergraduate research experiences, courses driven by inquiry or open-ended problem-solving, capstone courses, internships or jobs with a substantial mathematical component, etc.

In this paper we describe the structure of the MIT Project Laboratory in Mathematics (Section 2) and instructor roles (Section 3). We then report on the demography of students taking the course (Section 4), the projects, and how the projects have been tailored to the course and student population (Section 5). In Section 6, we review student reception of the Project Lab, and we end in Section 7 with some comments about transferability of this type of course to other educational environments.

This paper expands on a talk by the second author at the conference "New Directions for Mathematics REUs" at Mount Holyoke College on June 21, 2013. Haynes would like to thank the organizers of this conference for providing such a stimulating venue for disseminating new ideas about and roles for undergraduate research. The first author's involvement in this report arose from her work as the lead architect of a new category of publications by MIT OpenCourseWare (OCW) called OCW Educator. This collection of webpages seeks to portray not so much *what* we teach at MIT (the traditional OCW focus) as *how* we teach; the target audience is educators. One of the first MIT subjects to receive in-depth exposure on OCW Educator was the Project Lab [7]. Kathy is particularly well prepared to report on this subject, having taken it as an undergraduate. She would like to thank her OCW co-worker Kimberly Li, who worked with her on publishing Project Lab on OCW. We both thank Susan Ruff of the MIT Writing Across the Curriculum Office. Susan has been a constant with this course over the years and has been instrumental in its development, especially the communication component and the workshops. We also thank Michel Goemans for providing us with much of the enrollment data we report on below, and Jeff Lagarias for providing information about his variation on this course at the University of Michigan.

2. Structure of the Project Laboratory

Unlike most research experiences for undergraduates, which typically take place outside of class and often in the summer, the Project Lab is a semester-long course that takes place during the school year. This enables students to experience mathematical research while satisfying a degree requirement, and without the commitment of a summer.

Each semester begins with an introductory class, a teamwork workshop, a presentation workshop, and a writing workshop.

Over the course of the 14-week semester, students work in teams of three on three open-ended research projects. The project cycles are deliberately overlapped to allow time for feedback and revision. Each project cycle lasts five to six weeks and includes the following:

- *Topic selection*: Each team selects a project topic from a list of over forty options. In most semesters each project topic can be chosen by at most one team.
- *Mathematical work*: Students work together in their teams to explore their project topics. Students may work out examples, run computer simulations, read relevant literature, refine or redefine the focus of the project, make conjectures, and attempt to construct proofs.
- *Mentor meetings*: For every project, each student team is assigned to a mathematics instructor who serves as the team's mentor. Each team meets with its mentor once a week for the duration of the project. The communication instructor for the course works with all of the teams as needed.
- *First draft*: Each team submits a first draft of the paper ten days before the final draft due date. The paper is then read and commented upon by the team's mentor and sometimes also the communication instructor, while the students begin work on their next project.
- *Debriefing meetings*: The team meets with a group including the lead instructor, the team's mentor, the team's mentor for the next project, and often the communication instructor. The team gives a brief presentation of their research, and the whole group discusses both the mathematics and the writing.
- *Final draft*: Students finish a project cycle by revising their paper following instructor feedback and submitting a final draft.

Requiring three projects in a single semester is a deliberate choice. This quick turn-around is consistent with the rhythm of students' experiences in their other courses, which often involve weekly homework assignments and monthly exams. Having three iterations also gives students fresh learning opportunities: to choose more appropriate projects, conduct better research, and write better papers.

Each team delivers a presentation to the class on one of their projects during the semester. The formal presentation is always preceded by a practice presentation, which is attended by course instructors and includes extensive feedback. The team then refines its presentation before presenting to the entire class.

Communication is an essential component of the course experience. Project Lab satisfies MIT's Communication Intensive in the Major (CI-M) requirement, which mandates that all undergraduates take at least two courses designated as CI-M. CI-M courses are described as follows:

> CI-M subjects (Communication Intensive in the Major) teach the specific forms of communication common to the field's professional and academic culture. Students may write in teams, prepare and present oral and visual research reports for different audiences, learn audience analysis and peer review, or go through the experience of proposing, writing, and extensively revising a professional journal article.

The focus on communication provides a wider variety of assessed skills than is typical in a mathematics course. It is widely recognized that writing improves understanding. Moreover, it can be argued that the communication skills learned in this and other CI-M courses are among the skills we offer training in that are most easily transferred to the typical career paths of our students.

3. Staffing the Project Lab

The Project Lab is a labor-intensive operation. We limit enrollment to 27 students; once we tried it with 36 and the burden on the lead instructor was too heavy. The course is typically staffed with a lead instructor (a faculty member), two co-instructors (typically graduate students or post-doctoral researchers), and a communication instructor. Collectively, the team spends about 450 man-hours over the semester on the course. This serious commitment of resources allows us to provide students with personal mentorship, with encouragement and support as they confront the uncertainties of research, and with feedback and guidance on their presentations and papers. This student-to-instructor ratio is roughly in line with that of many of MIT's other upper-level mathematics courses.

For each project, each course instructor mentors three of the nine student groups. The mentor typically meets with each group once a week. These meetings are similar to meetings that a faculty member might have with a graduate student. Often, mentors do not know any more about the project than the students do, and they think through the issues together. Instructors try to be honest and sincere, and not feign ignorance. It is important that mentors try to help the students while at the same time giving them enough leeway to find their own direction, since the whole point of the course is for students to experience research themselves. If an instructor dominates the conversation or sets the research direction, the instructor can rob the students of the research experience. The meetings with mentors and the debriefings are designed to provide timely, formative feedback that encourages learning and spurs continued progress.

At the same time, most students are truly novice researchers. Sometimes they need help understanding the research process, identifying ways to tackle a problem, and feeling comfortable defining a research direction. Instructors help students cope with the challenge and uncertainty of confronting a question that does not have a known answer. Sometimes a week goes by with no visible progress, and instructors might offer a gentle push in some direction. Instructors try to be sensitive to students' worries, encourage the students, and help them overcome obstacles.

After a team submits a first draft of a paper, the mentor and sometimes the faculty lead instructor and the communication instructor read and mark up the draft. Then a "debriefing" meeting is held. This meeting is attended by the students on the team, their mentor, the lead instructor, and sometimes the communication instructor and the mentor who will work with that team on the next project. At this meeting, the students give an informal description of their research, and then everyone discusses both the mathematics and the writing. This meeting serves several purposes. It provides students with some practice at presenting mathematical ideas. It serves as an opportunity for the instructors and students to discuss the students' research and explore possible ways to push the students' work further or in new directions. And it gives the instructors a chance to talk through improvements for the paper rather than depending entirely upon written commentary.

The students as well as the instructors can appreciate this close working relationship. Many students have never worked with a faculty member, postdoc, or graduate student this closely before, and the course can give them a personal connection to all three. For the instructors, one of the most enjoyable aspects of the course can be getting to know a particular group of students very well.

The course lead is always a faculty member, who, along with our experienced writing instructor, offers guidance and feedback to the other two instructors. The course thus provides an excellent and exciting training ground for the mentors who are less experienced postdocs or graduate students. In fact we feel that the qualities of good mentorship found, for example, in [6], provide good orientation for the course instructors. Their INSPIRE model suggests that the mentor should be: Intelligent, Nurturant, Socratic (not didactic), Progressive, Indirect, Reflective, and Encouraging.

Educational initiatives stand or fall on how well they are adopted by individuals other than the initiators. On these grounds the Project Lab has been a signal success. It has never been difficult to staff. The following twelve faculty members have led the course, aided by quite a large number of postdocs and graduate students.

Mike Artin, Spring 2004, Spring 2006, Fall 2006
Haynes Miller, Spring 2005, Spring 2013
Bjorn Poonen, Fall 2007
Paul Seidel, Spring 2008, Spring 2009, Fall 2009
David Vogan and James McKernan, Fall 2008
David Jerison, Spring 2010, Spring 2014
JuLee Kim, Fall 2010, Fall 2011
Scott Sheffield, Spring 2011
Tomasz Mrowka, Spring 2012
Richard Stanley, Fall 2012
Larry Guth, Fall 2013

In each case, the lead faculty made additions or improvements to the project list and brought new ideas about workshops or other elements of the course process and grading method.

A certain level of constancy of staffing is also beneficial, and in the case of the Project Lab this has been provided by Susan Ruff of the MIT Writing Across the Curriculum office. Susan has worked with the course almost since its beginning. She knows the history of the course and what has or has not worked, and she helps faculty who are new to the course understand and structure the course. We have also been helped by the MIT Mathematics CI Space, designed by Sami Assaf, Mia Minnes, and Susan Ruff, and maintained by Susan. This is an instance of the Educational Collaboration Space, a publicly available WordPress package found at ecs.mit.edu. We have used it since 2010 to harvest and disseminate ideas and course material pertaining to the communication-intensive courses in the Mathematics Department. This resource, and most of the MIT material, is now publicly available through the MAA-supported website mathcomm.org.

4. Student Demography

The Project Lab is offered every semester, and it is open to all MIT students. Because the course is the only mathematics course that fulfills the MIT Undergraduate Laboratory Requirement, and it also fulfills one of MIT's communication requirements, the demand for the course is generally high; usually 40 to 50 students register to take the course each semester, well over the cap of 27 students.

Regrettably, the enrollment cap effectively limits the course to mathematics majors and to upperclassmen. Only 12 of the 489 students who completed the course by spring 2014 were not mathematics majors. On average, 72% of 18.821 students are seniors when they take the course, and 26% are juniors. While it has not been possible to accommodate many students from outside the Mathematics Department, it has been possible to guarantee that all mathematics majors can take the course if they choose to use it to fulfill their Laboratory Requirement. The drop rate has been extraordinarily low: about 1.3%.

The only enrollment requirement for students is that they first complete at least two mathematics courses beyond the basic courses in calculus, differential equations, and linear algebra. Many students enter the course with no substantial research experience, while others enter the course having already published papers in professional journals. This course is intended to be broadly accessible, within the MIT context.

Mathematics majors at MIT go on to a huge diversity of careers. About 40% continue after graduation in some academic pursuit, and 50% take jobs of various types. From the first group, about a third enter an engineering masters degree program, a third go into a Mathematics PhD program, and the rest go into PhD programs in some other discipline. Among the non-academic career group, about half go into investment banking and financial services; consulting and software development attract about 15% each; and the rest move on into a wide range of other activities.

A cross-section of mathematics majors do take the course, as indicated by the following statistics (from the past fourteen semesters). Many MIT undergraduates are "double majors," and the Mathematics department hosts more second majors than any other department: over the past seven years about 11% of our majors have listed Mathematics as their second major. Many other majors at MIT require a laboratory subject in the department as a graduation requirement, but nonetheless fully 19% of Project Lab students list Mathematics as their second major. Among them, more than a third are Electrical Engineering and Computer Science majors, more than a quarter are Physics majors, and the rest are scattered over ten other majors, tracking closely with overall trends.

Institute grade point average and rank within the Mathematics major for the Project Lab students closely matches the departmental averages. The mean GPA of majors graduating over the past decade is 4.53/5.00; among the mathematics majors who took the Project Lab it is 4.49/5.00. An arcane departmental ranking system uses grades and a measure of the difficulty of the class to assign a score to each senior, ranging, in the class of 2014, from a top score of 82 down to the low of -8, with a mean of 21.0 and standard deviation of 19.7. The mean score of Project Lab students was 23.6, one-seventh of a standard deviation above the mean.

Mathematics majors at MIT can choose from a wide variety of mathematics courses; they can elect to specialize their program as they like. In order to understand whether the Project Lab attracts students from one or another corner of the major, we determined, for each undergraduate subject (except ones carrying Communication Intensive credit, which are subject to other pressures) the number of students graduating with a mathematics degree by May, 2014, who passed that subject, and the number passing that subject who also took the Project Lab. We then aggregated these data by summing over courses in similar areas, and took the ratio of the resulting sums. This gives a crude indication of whether the course is more popular among some groups than others. The table below presents this summary. So for example 56% of the seats occupied by undergraduate mathematics majors in analysis subjects were occupied by students who by the time they graduated had also taken the Project Lab.

Analysis	56%
Topology, geometry	53%
Algebra	50%
Combinatorics, physical appl. math.	47%
Computer science, probability	42%

Students taking classes in the last category availed themselves of the Project Lab about 75% as often as those taking classes in the first category. This somewhat lower participation of students interested in combinatorics and computer science is at least partly accounted for by the large course 18.310 Principles of Discrete Applied Mathematics, which also carries communication-intensive credit, and the availability of Lab courses in computer science.

Grading a class like this presents challenges to those used to basing a grade on numerical exam and homework scores. Some thoughts about this can be found at the OCW Educator website [7]. For now we just report on the observed grade distribution. In the semesters since fall, 2006, the grades have been distributed as follows (percentage):

A+	A	A−	B+	B	B−	C+	C	C−	D
5	31	20	14	18	7	2	2	1	1

This distribution is somewhat higher than the overall departmental grade distribution, but we feel that generally students devote themselves to this course and these grades have seemed fair to us.

5. About the Projects

The Project Lab is taken by 54 students per year, with students' backgrounds and interests covering a broad span. In many research programs for undergraduates, the end-goal is novel, publishable results. This is not the case with the Project Lab. Instead, we aim for a "research-like" experience. There is no expectation that the teams discover previously unknown mathematics, though sometimes they do; the key is that they discover mathematics previously unknown *to them*, through a research-like process. Thus the Project Lab is a mathematics course, not what mathematics educators would typically call an "undergraduate research experience." This goal makes the experience accessible to students who do not yet have enough mathematical background to work at the forefront of mathematical research. It also enables the course to be repeatable, since projects can be reused from year to year.

The project descriptions have been crafted with this perspective in mind. Students begin each project by choosing from a list of over forty projects that has been developed by the mathematics faculty over the past decade. Each project description presents an open-ended mathematical situation, suggests some relevant questions, and allows students to define and pursue a range of research directions. The success of the course depends upon a list of good project topics.

While we prefer not to publicize the project descriptions, a couple of examples can be found at the OpenCourseWare website [7], and the second author invites the reader to contact him for more information.

The following are characteristics that we have found to be important features of a good project topic:

- *Open-ended, with a variety of possible research directions.* In mathematical research, we rarely aim at one specific target. Instead, we look at a general area and try to find the parts of it that are both interesting and accessible. We want students to experience the challenge of defining and pursuing their own specific research directions.
- *Accessible to students across a spectrum of backgrounds, interests, and research abilities.* The projects span pure mathematics, applied mathematics, and combinatorics. It is important that the projects appeal to students with different levels of preparation or areas of expertise. Many projects work as well with less well-prepared groups as they do with very well-prepared groups. It has been quite remarkable to see what different teams have done with the same projects and to appreciate the diversity of results that students have produced. This brings home the message that mathematics has a broad entry path, which people of many different backgrounds can access. We feel that this lesson is an important component of the course. Successful papers and presentations require a substantial investment of time and energy, for everyone. This course, with its very heterogeneous enrollment, harmonizes with the sense that too much is made of intrinsic genius and not enough of hard work and motivation [3].
- *Manageable in a one-month period.* We want the students to have a taste of what it is like to do mathematical research, but we only have about one month for each project, and students often enter the course with little to no research experience. In contrast, professional mathematicians often spend months or years on a project and submit a paper only after substantial progress has been made. Our research topics are open-ended but focused enough that we believe most teams can make reasonable progress within the short timeframe.
- *Difficult to resolve via external resources.* We have found that absorbing existing research can be distracting or overly time-consuming for students, and we recommend that they focus

on exploring the mathematics themselves instead of diving into a full-blown literature search. Most of our topics are not associated with extensive literature and are not easily searchable on the web. When there is a giveaway term, we often rephrase and disguise the problem in a way that makes it much harder to find. It is fine if students find existing research on a topic, but they are then expected to build upon what they find or pursue a different direction.

The fact that student teams choose their own projects gives them a sense of ownership over their work. Moreover, the projects are conceived to allow a variety of interpretations, and the students take them in their own remarkably diverse directions. This sense of ownership is critical in motivating students to push on through the research, and especially through the writing. One of the big challenges in teaching technical writing is finding subjects that the learner is committed to; if the subject is the student's own original idea and work, the commitment is much easier.

Every student is expected to make progress that is commensurate with his or her background and experience. Some students barely make any progress on their first project but eventually become more confident and capable at approaching research problems and at describing their findings. Because the course is so open-ended, students can find a project that is accessible to them and define a sub-question or attempt an approach within their grasp. Our most accomplished students can learn to be more patient and improve their teamwork skills. They can gain experience tackling new research projects, and sometimes they find brilliant solutions. Many projects lead right to the frontier of current research, and students gain an understanding of the effect of deeper mathematics on these fairly naïve questions. These are all important experiences for budding mathematicians.

Education theory offers an interesting perspective on students' experience in the Project Lab. The course provides a context in which the student can engage in a form of "deliberate practice," described by Ericsson *et al.* [4] in these terms: The subjects experience

> "...motivation to attend to the task and exert effort to improve their performance. In addition, the design of the task should take into account

> the preexisting knowledge of the learners so that the task can be correctly understood after a brief period of instruction. The subjects should receive informative feedback and knowledge of results of their performance. The subjects should repeatedly perform the same or similar tasks."

The learning process in the Project Lab conforms well to the constructivist paradigm (springing from the work of Piaget), as described for example by Micheline Chi [2, pp. 85–86]:

> "Constructive activities ... allow the learners to infer new ideas, new insights, new conclusions, from making deductions and inductions, from reasoning analogically through comparisons, from integrating new knowledge with old knowledge, or linking information from disparate sources. In short, these various creating processes of comparing, connecting, inducing, analogizing, generalizing, etc., allow the learners not only to infer new knowledge but also to repair and improve their existing knowledge. How would these creating processes enhance learning? Inferring new relations, new conclusions, and new insights obviously makes one's knowledge more rich, and repairing one's knowledge also makes it more coherent, more accurate, and better-structured, and so forth. These changes can deepen one's understanding of new materials and have been shown to improve learning."

Chi refines the constructivist hypothesis in an interesting way, which seems relevant to the Project Lab. She distinguishes between the terms "active," "constructive," and "interactive." Active processes are likely to engage students more than purely passive ones, but may still bypass the brain cells. Constructive activities require more engagement, as illustrated in the quote above. But activities which respond to and stimulate a response from the learner's environment form a separate category. Chi describes several varieties of interactive process, including joint dialogues with peers.

> "When a learner interacts with a peer, such interactions can sometimes characterize a pattern of *joint dialogues*, which occur when both peers make substantive contributions to the topic or concept under discussion, such as by building on each other's contributions, defending or arguing a position, challenging and criticizing each other on the same concept or point, asking and answering each other's questions. ... Thus *joint dialogues* refer to a pattern of interactions in which both partners make substantive contributions to the topic or concept under discussion ...."

The power of this form of knowledge-building is part of what makes the teamwork aspect of the Project Lab so important.

6. Student Reception

We have tracked student attitudes quite closely. The following data are assembled from responses to the spring 2013 departmental opinion survey of our graduating seniors. One question we ask is: "What Mathematics course did you find most beneficial?" Given the non-syllabus nature of the Project Lab, it is perhaps surprising that Project Lab ranks third in the rate of mention (mentions per student taking the class), with six of the 62 respondents mentioning it. The top two are Algebraic Combinatorics and Theory of Computation. So the Project Lab ranks above [Modern] Algebra I, Real Analysis, and Probability and Random Variables (which come next in this ranking). It was also very interesting to discover that five of these six students were not bound for an academic mathematics career; in fact only two of them were going on in academics at all.

Here are some student comments mentioning the Project Lab as the mathematics course they benefited from the most, followed by the student's post-graduation plan.

- "It gave me a lot of practice in working on a challenging problem and writing and presenting about it. It definitely got me thinking about how to communicate about my work, much more so than the seminars I took." (Statistics PhD)
- "The most beneficial class that I have taken is [the Project Lab]. Although my teammates and I struggled a lot when working on the papers, we learned how to approach an open-ended problem and discovered new problems." (Financial sector)
- "This was the first time I had to give a formal presentation and write a formal math paper. These skills are invaluable and I have used them countless times since taking that class." (Software)
- "Working in a group, working on writing and papers, more faculty interaction than other classes. It was good!" (Consulting)

We also ask students about their experiences with our "Communication Intensive-Major" offerings. In response to "What is the best thing about the Mathematics CI-M program?", the Project Lab was mentioned almost twice as often as any other subject. Here are some comments in reaction to this question, followed by the student's post-graduation plan.

- "The chance to present mathematics and receive feedback on presentations." (Mathematics PhD)
- "I thought the Project Lab was fun, the problems were just the right level, so they were challenging but not impossible." (Mathematics PhD)
- "The Project Lab was a great course, as it involved both original work and communicating about it." (Non-mathematics PhD)
- "I liked working on Math Projects in the Project Lab. It was very interesting and the scope of the projects was about right." (Software)
- "Giving presentations — I learned a lot from them. [X] was an awesome teacher; he put a lot of time into meeting with our group, and really thought about our project at the board alongside us. I wished we could have more experiences like this!" (Non-mathematics PhD)
- "Learning to give long lectures/talks was great. I really loved the problems presented in the Project Lab, we had to work collaboratively and produce results in a fixed amount of time." (Software)
- "[The Project Lab] was great. Faculty interaction, good setup to learn writing, group work, open-ended problems." (Consulting)
- "For the Project Lab, working a long time on a particular problem was fun, especially since we were able to pick problems that were particularly interesting to us. I was also working with my roommate and a close friend so that's always nice." (Software)

In response to "What is the worst thing about the Mathematics CI-M program" there was just one complaint about the Project Lab: "A specific project advisor in project lab was awful to deal with." (Mathematics PhD).

The MIT end of term subject survey also always generates quite a bit of commentary by students, on all aspects of the course. A principal theme of these comments is the student appreciation of the mentorship they received in the course. Here are two interesting responses reflecting this, and exhibiting the self-directed learning that takes place.

> "This has been one of the most memorable classes I've taken at MIT. It was more work than I initially imagined, but very gratifying to be able to write 3 full papers and give a full-length presentation. The amount of feedback we received on all assignments was so detailed and so helpful, and I have definitely grown as a writer and presenter. Thank you very much."

The next quote responds to a request for commentary on the projects. We mention it not to highlight one or another of the projects, but rather to illustrate the student control over the learning process that this course can engender.

> "Project 1: I knew from past experience that this question would be painful to tackle. However, once proper questions (as Professor ...repeatedly notes) were carefully mathematically put, either results would follow immediately or the question transforms into a much deeper one. [Our mentor] was very helpful in guiding us on whether the questions we were asking were properly put and if our formulation would just lead to further complications (along with other guidance). Although I got results with the basic algebraic approach, I went on to investigate for a long time the more abstract algebraic approach. So, I learned from this first project the important skill of asking questions properly, but I did not learn the important skill of discarding complicating formulations early on. Project 2: What I learned and what I didn't learn from the first project carried on here. A problem that arose here was that I thought the natural way to look at this project was the algebraic one. However, I should have discarded this approach early on for the sake of time. Project 3: By this time in the semester, I've mastered the second skill I mentioned above, namely, that I should discard complicating paths early on. This skill enabled me to explore different ways to look at this problem in the time allowed."

Practice presentations turn out to be a very important and popular part of the course. Here are some comments reflecting that.

- "Most valuable part of the class."
- "Learning to give mathematics presentations is a different skill than that of giving general talks. I needed this skill, and the presentation part of the course gave me this skill. It was very helpful

to have one and only one practice presentation. That way, we get prepared and very prepared early. The comments were very helpful and applicable. The presentation experience was very nice."

- "The practice presentation was an incredibly useful couple hours for us because we had the wrong approach and were not aware of it. We worked hard on improving our presentation and it felt very satisfying when we finally presented, delivering what seemed like a clear message (based on students' comments)."

Other student comments have centered on two issues: It is tricky to communicate the grading rubric to students, and in some semesters there have been significant complaints about this. Teamwork can be problematic, and students whose teams become dysfunctional were not shy about reporting the resulting problems.

7. Other Instances of the Project Lab

Variations on this course have been offered elsewhere. The course Math 389 Explorations in Mathematics has been offered at the University of Michigan for a number of years under the direction of Professor Jeff Lagarias, as part of the Mathematics Department's Center for Inquiry Based Learning. The problem list began as a copy of the MIT list, but has evolved somewhat differently. There are interesting differences between the UM and MIT courses. The UM course is regarded as a toolkit course, training in the nuts and bolts of constructing and writing up proofs. It is addressed to students early in their undergraduate career. Lab space is provided, where students can find a mentor. There are two practice presentations rather than one, and each team presents twice in the course of the semester. It enjoys an enrollment of about a dozen student each year, despite fulfilling no institutional requirements.

Another version of the course was offered for a few years at the University of California, Berkeley, again using the MIT problem list.

We feel that courses of this type provide important elements of an undergraduate education in mathematics not well addressed by traditional curricular courses. The OpenCourseWare website [7] is designed to provide guidance to others interested in setting up such a course.

References

[1] S.A. Ambrose, M.W. Bridges, M. DiPietro, M.C. Lovett, M.K. Norman, *How Learning Works.* Jossey-Bass, 2010.

[2] M. Chi, Active-Constructive-Interactive: A Conceptual Framework for Differentiating Learning Activities, *Topics Cognitive Sci.,* **1**, (2009) 73–105.

[3] G. Colvin, *Talent is Overrated.* The Penguin Group, 2008.

[4] K.A. Ericsson, R.T. Krampe, C. Tesch-Romer, The Role of Deliberate Practice in the Acquisition of Expert Performance, *Psychol. Rev.*, **100**, (1993) 363–406.

[5] S. Greenwald, H.R. Miller, Computer-Assisted Explorations in Mathematics: Pedagogical Adaptations Across the Atlantic, in *University Collaboration for Innovation: Lessons from the Cambridge-MIT Institute.* Sense Publishers, 2007, pp. 121–131.

[6] M.R. Lepper, M. Dreake, T.M. O'Donnell-Johnson, Scaffolding Techniques of Expert Human Tutors, in K. Hogan and M. Pressley (eds.), *Scaffolding Student Learning: Instructional Approaches and Issues.* New York: Brookline Books, pp. 108–144.

[7] H. Miller, N. Stapleton, S. Glasman, S. Ruff, *18.821 Project Laboratory in Mathematics.* Spring, 2013. (MIT OpenCourseWare: Massachusetts Institute of Technology), http://ocw.mit.edu/courses/mathematics/18-821-project-laboratory-in-mathematics-spring-2013 (accessed on 27 September, 2014).

[8] R.D. Roscoe, and M. Chi, Tutor Learning: The Role of Explaining and Responding to Questions, *Instr. Sci.*, **36**, (2008) 321–350.

Chapter 4

REUs with Limited Faculty Involvement, "Underrepresented" Subjects in the Undergraduate Curriculum, and the Culture of Mathematics

Yanir A. Rubinstein
University of Maryland, College Park, MD, USA
yanir@umd.edu

Ravi Vakil
Stanford University, Stanford, CA, USA
vakil@math.stanford.edu

Research Experiences for Undergraduates (REUs) in mathematics started as an experiment in the 1980s. As described in the introductory chapter to this volume, the handful of programs established back then were solicited by an experimental National Science Foundation (NSF) program, and partly as a result of that, the structure of these different REUs was rather homogeneous. Of course, the programs were quite different in their offerings, and perhaps even in their philosophies. But the main structure was more or less set in stone: a faculty member (usually one) leading a small group of undergraduates during about two months in the summer in a research topic deemed "accessible" to an undergraduate entering his/her junior or senior year. The faculty member was expected to be around for that whole period and devote the bulk of his or her time to the project.

MSC 2010: 01A67, 01A80 (Primary); 58-03 (Secondary)

This model has been functioning with rather astonishing success over the past quarter of a century, and has had a profound influence at all levels on mathematics as well as on the *culture of mathematics.* We make some comments on this latter topic in Sec. 5.

Over the years, this model has evolved in many directions, some of which are described in this volume. In this article, we do not wish to challenge this initial model. (What is more, one of the authors is an alumnus of this model.) Rather, we wish to present thoughts on how REUs can evolve in different "ecosystems" where the standard model has not taken root. As an example, we will present some preliminary results on an experiment we have been part of during the past three years at Stanford University. This involved trying out different models for REUs, as well as some experimentation with the research topics chosen. We then reflect on the successes and failures of our approach, as well as on some related "philosophical" thoughts related to the future and culture of mathematics.

The danger of having a "standard model", no matter how successful, is that it provides reasons for not doing something. These are the reasons some colleges/universities, in the authors' experience, have been reluctant to have activities related to undergraduate research in mathematics. "I didn't do this as an undergraduate." "I don't think undergraduates can contribute significantly to cutting-edge research." "I don't think that's the best model." "That model requires more faculty commitment than we can afford." "We don't have the resources to do that." "I can't see myself being [famous successful mentor name]." Instead, we should figure out what we as a community want to get out of REUs and what we want the students to get out of it.

We also argue that REUs must have value for the other people (i.e., the people other than the participants themselves) involved, or else it will not happen (or at the least, will not be sustainable). Service is seen by some in our community as something with negative connotations, without recognition, to be avoided. To be specific, in some institutions, such service does not "count" toward the usual responsibilities of the faculty member. Mentoring in REU should not be service in this negative sense of the word (from the point of view of the mathematical community), although obviously it is service in the positive sense of the word.

The case we wish to make is that the mathematics community should not fail to create valuable opportunities for students because we cannot always have a "perfect" REU. On a slight variation on the words of Voltaire ("...le mieux est l'ennemi du bien"), the perfect is the enemy of the good. Instead, we should make the most of the resources available.

1. The Stanford Undergraduate Research Institute in Mathematics (SURIM)

The Stanford Undergraduate Research Institute in Mathematics (SURIM) was founded in 2012, based on a proposal drafted by R.V. (Ravi Vakil) and funded by the Office of the Vice Provost for Undergraduate Education and the Department of Mathematics.

Prior to SURIM's inception, undergraduate research at Stanford was limited to one (faculty)-on-one (student) (occasionally, one-on-two) summer projects, in a typical year limited to a handful of students. The limitation mainly came from the fact that only so many faculty could devote a considerable portion of their time during the summer to such endeavors (having many other conference and research obligations during that time). However, over the years it became rather clear to those in close touch with the math majors community that there was a strong demand for more research opportunities. For one, admission to REU programs around the continent became increasingly tough. To quantify this demand, a quick survey of math majors revealed that there were around 35 math majors interested in pursuing research over the summer. Convinced by these numbers, the Department's administration relented in its initial resistance to the program, which paved the way to its inauguration.

As just described, SURIM was initiated mainly in order to quench the thirst of Stanford's math majors for more research opportunities. At the same time, given this opportunity, we have consciously tried to use it to experiment with different ideas in the realm of undergraduate research. In the next subsections we describe in detail the SURIM program, and in Sections 2 and 3 we touch on some of these latter aspects.

1.1. Funding

Prior to 2012, the Office of the Vice Provost for Undergraduate Education typically funded about 3–6 undergraduates each summer in one-on-one research with a faculty member. (Many more students were potentially interested in taking part in some sort of undergraduate research experience, but could not find a Stanford faculty member willing and able to supervise them. Some went elsewhere for REUs, and many did less mathematical things with their summers.) Starting with SURIM's inception in 2012, funding was made available for up to 12 students. By giving course credit to a few more students, the program's scope was increased to 15 students; namely, three additional students were not given stipends but instead were given course credit. The University was supportive of the Mathematics Department's initiative of SURIM, and was happy to fund it at a higher level than it had funded the department previously (as described in the opening sentence to this subsection).

1.2. Framework

SURIM was a ten-week program during the Summer quarter (typically starting in late June) that ran in the three Summers of 2012, 2013, and 2014. About 12–15 undergraduates from Stanford took part in it each summer. Of these, about 3–5 worked one-on-one with a faculty member, while the rest were divided into groups of 3–5 students and worked on collaborative projects. The different models for the research work carried out in SURIM over the past three years are described in Section 2. In this subsection we concentrate on the general framework and format of the program that was common for all the different groups (by groups we refer both to the teams as well as the one-person teams consisting of the individual students). We end this subsection with a description of the administrative structure of the program.

Each group had several formal meetings with mentors each week, typically three times a week. The first two weeks' meetings were mostly devoted to lectures by the mentors, introducing the problems to be tackled and the basic circle of ideas surrounding them. Jointly with the groups, the mentors decided on how to divide the work, and which problems to concentrate on. Members of each group then prepared tutorials to other group members, as well as presentations to the entire institute each week, giving a status report to those working on other problems. Much of the rest of the week was spent by participants working individually and in groups, and in informal discussions with mentors. The last 2–3 weeks were mostly devoted to preparing a report and a final presentation to the entire institute, to which also the entire department was invited. (An unexpectedly large number of faculty attended the day of presentations, and in turn the faculty were surprised at how interesting the students' results were.)

Several additional events were scheduled each week, of which we describe a few representative ones. First, there were introductions to LaTeX, as well as to various software packages (such as Matlab). In addition, most weeks featured a guest lecture usually from a faculty or graduate student, as well as one or two presentations from mathematicians working in industry. The presentations also touched sometimes on career choices — working in academia or industry, and what mathematics research can be about. A one-time event that has been featured every year is the graduate school panel, where graduate students present their experiences on applying to graduate schools, career choices, and answer questions on these and other topics. Another event that took place in 2013 was a joint meeting with the Berkeley RTG REU program and presentations (for more on this we refer to Cristofaro-Gardiner's article in this volume).

An activity that was launched based on input from participants was the "take a professor for lunch" initiative that proved to be quite popular. Here, each group invited a faculty member of their choice to an informal lunch in a relaxed setting.

1.2.1. *Administrative structure*

We end this subsection with a few words on the administrative structure of SURIM. Two tenured faculty were in overall charge of the program (Gunnar Carlsson and R.V.). Their role was mostly in overseeing the program and hiring a Managing Director (MD) for the program. In the first two years (2012, 2013) the program MD was Dr. Nancy Rodriguez, then a postdoc at the Mathematics Department at Stanford (and now on the faculty of North Carolina State University).

In 2014, because Dr. Rodriguez had received an offer from NCSU, the MD was Chris Henderson, then a graduate student in his penultimate year in graduate school (who had served as a graduate mentor in SURIM both in 2012 and in 2013). The MD's role was to oversee the day-to-day activities of SURIM, for instance: organize the guest lectures, recruit the graduate student mentors, organize the applications and admissions, manage the program's webpage. The MD's salary was funded by the Department of Mathematics, as was that of the graduate mentors. (The support for the graduate students was modest compared to the work done.) What is more, some of the graduate students seemed to devote much more time than what was formally required of them. We come back to this in Section 2.

1.3. Applications and admission

Because the funding came from Stanford's Office of the Vice Provost for Undergraduate Education, SURIM was only open to Stanford students. On the other hand, there were no restrictions on citizenship.

Key to SURIM's success was active advertisement and recruitment. All mathematics majors, and all members of the Stanford Undergraduate Mathematics Organization were emailed. Signs were posted prominently in the building. Professors mentioned it in key classes. Some professors approached mathematically talented freshmen from underrepresented groups, to encourage them to apply; such direct recruitment was particularly effective.

One-third of the applicants received first-round offers. In the first year, the organizers realized that offers had to come out early enough to be competitive with internships and other summer research experiences. The support over the summer (in terms of duration and amount) had to be in line with what other departments were able to offer. The support in summer 2012 was $4,800.

Selection was made by the organizers, as well as by the mentors who would be spending the most time with the students.

2. Three Experimental Models

SURIM experimented with several models for undergraduate research. For example, in 2012, there were a handful of students working one-on-one with a faculty member (the traditional model, described in Section 2.1); there was a group of three students working with a faculty member following an experimental model, the main topic of this section, described in Section 2.2; and, finally, there were two groups of 3–6 students each, mentored solely by graduate students, see Section 2.3.

2.1. One-on-one research with faculty

This was the traditional model available at Stanford before SURIM. Students would not work in teams (with the exception of students working in Gunnar Carlsson's large applied topology group).

2.2. Part-time faculty involvement

Math REUs have traditionally been hosted in non-R1 institutions (by R1 we refer to the class of Research I institutions, a terminology used in the Carnegie Classification of Institutions of Higher Education). For example, based on an outdated list of R1 institutions [6], and a list of the current NSF funded REU sites in the Mathematical Sciences [7], 36 sites are currently in non-R1 sites, while 10 are in R1 sites (i.e., about 22%).

What are the reasons for such a distribution? In particular, why are so few REUs held at R1 universities? This seems paradoxical, precisely since such institutions typically have better research resources and infrastructure available.

We do not attempt a comprehensive answer to this question. However, we believe that one of the major reasons for this is that most R1 faculty have quite limited time available to conduct their own research-related obligations (which comprise a major component of their job duties), and the summer time is mostly dedicated to these responsibilities. Indeed, the traditional model of REUs developed in the late 80's is based on full time faculty involvement. That is perhaps one of the main reasons that such models have been adopted by so few R1 universities.

We now describe an attempt to test out a possible model that might address this point. This model was experimented with by Y.A.R. (Yanir A. Rubinstein) in 2012 during the inaugural year of SURIM, and we later describe the results of that experiment.

2.2.1. *The model*

In this model, a group of 2–6 undergraduates is mentored by a faculty member (FM), with the heavy assistance of a graduate student (GS). We emphasize that the faculty member could be at any stage of the career: postdoc, tenure-track or tenured. In fact, the fact that some R1 institutions have postdocs (as opposed to non-R1 institutions) perhaps allows a novel way to involve young faculty in REUs, since oftentimes postdocs at these institutions have less administrative/advising (and often even less teaching) obligations. The GS might be already working with the FM, but could also be in early stages of his/her studies. In the latter case, the GS might be possibly interested to learn more about the FM's research before choosing a thesis advisor, for which this work experience could be quite vital.

The main responsibility of the FM is, above all, to suggest the research problems for the group. As usual, this would require quite some preparation. Once a reasonable pool of problems is prepared, the FM lays out a road map for the summer, with input from the GS.

Within this model, both the FM and GS are expected to share the most time-consuming responsibility during the summer, that is, the formal and informal meetings with the students.

First, perhaps differently than most traditional REUs, the meetings are limited to three per week during most of the summer, and reduced to twice a week occasionally (perhaps during 1–3 of the weeks). The purpose of this is twofold. On the one hand, it is meant to give the students more responsibility and freedom in their work, and a chance to explore research avenues on their own. The reduction to two meetings a week typically occurs during weeks of intense writeup or preparations for the final presentation, but are also based on adapting to the FM and GS's travel schedule. One of our students commented at the end of SURIM about this point: "The meetings are spaced out enough to allow students enough freedom and let them work at their own pace, but they're not so far as to let them procrastinate. I think this is a real problem with the summer. Had we been meeting only once a week, then there would've been a greater tendency to work on days 4-5-6." On the other hand, it is meant to allow both the FM and GS to devote time to other matters, such as family or research.

Second, while during the first two weeks both FM and GS are expected to be around to get the project on its feet, that is not the case for the rest of the remaining 6–8 weeks. During those remaining 6–8 weeks, the FM is expected to be in residence for about half of the time. In addition, the GS is expected to be present for about two-thirds of the time. Of course, the FM and GS must coordinate their travel appropriately. Thus, assuming a typical REU duration period of 8–10 weeks, the FM is typically around for 4–6 weeks, while the GS for about 5–8 weeks.

2.2.2. *Possible advantages*

While the students are not able to benefit from *full time* interaction with the FM, there are several advantages to this model.

The first and perhaps not the most obvious one is the ability of the three key players (FM, GS, students) to experience interactions happening at other levels. To be specific, the students profit from being present during high level interaction between the FM and the GS, giving them a glimpse to graduate school and the joys, frustrations, and responsibilities of a research mathematician. The GS gets a glimpse, as well as real experience, in the art of advising students. The FM gets a glimpse at the real difficulties the students are facing via their interaction with the GS.

Another advantage is the fact that the project is not a full time commitment for the FM or the GS (although it is of course a "most time" commitment). This allows strong and active members of the faculty and graduate student body to commit to such an activity, which, of course, ultimately benefits the students.

Moreover, the GS benefits from a possible additional research avenue, as well as advising experience and, sometimes, additional publications. These ultimately could be very helpful in career development.

What is more, the less than intense meeting schedule (i.e., only 2–3 times a week) seems to contribute to a strong collaboration and exchange of ideas between the students who gain additional freedom and must rely more on each other for learning some of the background.

Last but certainly not least, this model allows the training of the next generation of research mathematicians that might be more open than otherwise to the free exchange of ideas and less competitive collaboration style, as well as encourage them, when the time comes, to foster in a similar manner younger mathematicians-to-be. We make additional comments on this subject in Section 5.

2.2.3. *Results of the experiment*

We now describe our experience with this model based on research interaction with a group of three undergrads led by Y.A.R. (the FM) in SURIM 2012.

The GS was a rising second year graduate student (Otis Chodosh). As a preparation to the REU, the FM taught a graduate topics course in the Winter quarter that the GS actively took part in. The course also allowed the FM the chance to collect a pool of possible research projects and write some lecture notes.

We now describe the temporal structure in terms of the presence of each of the mentors. During the first two weeks of SURIM, the FM used some of the lecture notes described in the previous paragraph to introduce to the group of students and to the GS the subject and propose a few possible problems. The FM was then at a conference for about 10 days, during which the GS met the students on a regular basis. When the FM returned, the GS left to a summer school for two weeks. Upon the GS's return, the FM was traveling for a week. The zigzag continued in this manner until the end of the program, with usually little physical overlap between the GS and FM. However, the students themselves updated the GS and the FM of their progress during their respective absences. Moreover, the GS and FM kept close contact during those times via phone and e-mail.

The group of students was deliberately the strongest that was recruited that year for SURIM. At the same time, two out of three of the students were rising sophomores with quite little background to prepare them to actual research. In fact, neither of these two students were among the first round offers for admission that year. Whether SURIM had a role or not, later these two students were considered among the most successful math majors in their class. One went on to graduate school at MIT, while the other to UC Berkeley. Moreover, the latter student completed his senior thesis under the supervision of the FM in 2015 ultimately winning the 2015 Kennedy prize for the single best undergraduate thesis in the natural sciences in Stanford University (this is the *highest honor* an undergraduate can achieve at Stanford). The third student was merely a rising junior, but served as a mentor of sorts to the other two in terms of background in more advanced topics. Unexpectedly, each of the three students brought different perspectives to the group effort, complementing each other.

Paralleling that, the FM and GS complemented each other in terms of contributions to the project. After suggesting the main problem and several lectures on it, the FM suggested splitting the work into a numerical/simulation aspect and a proof-based aspect. The former was supposed, naturally, to help produce provable conjectures for the latter. At that point, the GS was very helpful in giving the students guidance into research methods (programming, literature search, etc.). As the summer progressed, the GS was also helpful in verifying some of the students arguments, finding small errors, suggesting alternative approaches, and so on. The FM mainly asked the students and the GS to explain to him what they have been working on, serving oftentimes as "an idiot in the room," or sometimes, as a devil's advocate — "why is this interesting?" At the final stages of the summer, both the FM and GS verified the correctness of the main result, and polished and slightly generalized it, and finally, guided the students towards the writeup of the final report by setting clear objectives and expectations.

Once the summer was over, the FM strongly encouraged and considerably helped the students to submit abstracts to the undergraduate research session in the MAA-AMS Joint Meetings. The FM also helped secure additional funding from Office of the Vice Provost for Undergraduate Education to fund the travel of the students to that conference (and used his own NSF grant to help fund his own travel and that of the GS). Two abstracts were accepted and successfully presented. Based on these, the students were further invited to an AMS Special Session for experts on the topic (*not* an undergraduate research session!). Finally, about 15 months after the program, the group (with a strong push from both the FM and the GS) wrote up a first article summarizing the main results. The article has been published since by a respectable peer-reviewed journal in the subject [1]. The revision process extended over Summer and Fall 2014 and provided further opportunity for interaction between the students, the GS, and the FM, as well as valuable experience for the students in the process of research publication.

2.3. Graduate-led undergraduate research at SURIM and other R1 institutions

Another model tried out extensively at SURIM was groups of undergraduates entirely led, mentored and supervised by graduate students. This model was not an exclusive innovation of SURIM, and we refer to Cristofaro-Gardiner's article [2] on this topic in this volume, concerning a similar model/experiment carried out at the same time (starting 2012) at UC Berkeley. Partly for this reason, and as the aforementioned article does a thorough job describing this model, we do not describe this model here in detail.

We make several general remarks instead. A variety of approaches to undergraduate research where most of the interaction has been with graduate students have been tried at other R1 institutions. For example, Berkeley [2], Chicago [4], MIT, and Harvard have active programs that anecdotally have been successful. We will not attempt a comparison of these programs. But any comparison of such programs with "identical" programs with graduate student time replaced by faculty time is silly — such resources (equivalent faculty time) are not readily available at these institutions. Furthermore, the graduate students involved were active researchers (albeit many just starting out), and were only a few years from becoming faculty members themselves. We expect that they will bring their experience and enthusiasm for working with undergraduates to their long-term institutions. In this way, such programs have long-term advantages for the community. Moreover, in the best of these graduate-led summer programs, the graduate students are well advised by faculty, and faculty are also directly involved with the undergraduates, even if not on a daily basis.

3. Underrepresented Areas in Undergraduate Research

Over the past 25 years, REU topics have concentrated mainly on research projects drawn out of fields in mathematics that seemed more easily accessible to US undergraduates completing their sophomore or junior year.

As a result, it seems plausible that certain areas of mathematics are "underrepresented" in the REU spectrum (as well as in the undergraduate mathematics curriculum — although we do not discuss this broader issue). This section does not attempt a scholarly study of this phenomenon. It is merely an attempt to raise this topic for further future discussion. In Section 3.1 we describe some "anecdotal data" that touches on this underrepresentation a bit more precisely, and perhaps gives some evidence for its existence. In Section 3.2 we describe an experiment one of us conducted in which an REU was designed in an apparently underrepresented area and in which students that had essentially no prior background in the field took part.

3.1. Some anecdotal data

In order to gain some feeling for the mix of subjects being offered in REUs, we collected data on past REU project topics. The data was collected by a student assistant as follows: starting with the NSF's list of currently available math REU sites, we searched those sites webpages for current and past REU research topics. Since only so many sites list their past projects, this resulted in only ~500 projects, that have almost all taken place in the past 15 years. Thus, while this data pool is significant, it is by no means exhaustive. The first column of Table 1 lists the very rough distribution of these 500 odd topics into subfields. Similarly, the second column lists the distribution of Miller's Mathematics Laboratory Project (MLP) topics at MIT [5]. Since the MLP topics were concentrated on pure mathematics, we have listed in the third column the "adjusted REU500" list, where we simply took out the applied mathematics and statistics projects from the first column. Finally, in the fourth column we list the distribution of subfields as represented by research articles that are reviewed by Mathematical Reviews, where the numbers here are very roughly derived from the precise data in [3]. (Of course, these numbers themselves could be argued upon, as in some fields there is a tendency to publish a smaller number of papers.)

Table 1 suggest that Analysis and Geometry are underrepresented in the sense described earlier, while Combinatorics, Number Theory, Algebraic Geometry, Algebra and Topology are "overrepresented".

3.2. An experimental REU involving analysis and differential geometry

In SURIM 2012, Y.A.R. led a group of three undergraduates in an REU project on Optimal Transportation (OT). From the organizational point of view, the research was structured as explained in Section 2.2, and, as described there, this structural model was itself an experiment. However, what we wish to focus on in this subsection are the mathematical aspects of this experiment. Namely, the project centered on aspects of OT that lie on the intersection of the study of partial differential equations (PDE) and differential geometry. This intersection is sometimes referred to as geometric analysis (which is one of the NSF DMS "Disciplinary Research Programs", alongside algebra and number theory, analysis, applied mathematics, combinatorics, computational mathematics, foundations, mathematical biology, probability, statistics, and topology). Traditionally, and according to Table 1, this area is underrepresented. In fact, this area is quite likely also underrepresented among the thesis topics of U.S. citizens obtaining a PhD in mathematics (although this is simply

Table 1. Anecdotal data on distribution of research topics (percentages).

	REU500	MIT MLP	AREU500	MR
Combinatorics	19	20	27	9
Number Theory & Algebraic Geometry	15	9	21	7
Algebra & Topology	28	10	38	12
Analysis & Geometry	8	33	11	20
Applied Mathematics & Statistics	28	0	0	45
Probability	1	17	1	3
Dynamical Systems	2	10	3	4

a hunch of the authors and not based on any hard data). It seems that the reigning philosophy is that this area, that is rather modern, requires a mathematical preparation and maturity that is rarely possible with undergraduate students.

Coming back to our experiment, as pointed out in Section 2.2.3, two of the team's students were merely rising sophomores, while the third was a rising junior. The younger students had never before studied PDEs, though they took honors calculus as well as real analysis classes, and one of them took an undergraduate differential geometry class. On the face of it, there seemed little hope to make much headway in the summer beyond perhaps a reading course introducing them to the very basics of OT, let alone approach a serious research problem.

However, as it turned out, geometric analysis can provide meaningful REU projects that can, in fact, lead to new results. Just as in combinatorics or number theory, it is possible to choose topics that involve challenging computer simulation that can lead to reasonable, and, sometimes, approachable, conjectures. The enthusiasm and fearlessness of such young investigators can actually cut through barriers that more experienced researchers might shy away from.

Even though the learning curve (for the students, but also, to a lesser extent, for the graduate student assistant) was rather steep in the first 3–4 weeks, once the computer simulation part of the project produced results (about midway through the summer), the theoretical aspect of the project took several leaps. The intuition gained through such hands-on intensive investigations gave the students an edge and made a difference towards proving interesting and new theoretical findings. Complementing these with guidance from the FM and the GS yielded a winning combination. As described in Section 2.2.3, the research led to invitations for talks in AMS sessions both on undergraduate research, but, much more impressive, also on applied PDEs, alongside established experts! Finally, one publication was accepted to a topical journal [1] (from the referee report: "The results are very interesting, and the referee is happy to have had the pleasure of reading the submission."), while at least one other paper is in the pipeline.

This is just one case study. Clearly, we are not claiming it is easy to design successful or meaningful REU projects in underrepresented areas. What is more, one could certainly argue that three undergraduates and one graduate student from Stanford do not represent the average REU team. Also, some luck certainly played a factor. All of these are true to some extent. Our purpose in this experiment, notwithstanding, was to make a small step towards shifting the stereotype surrounding the underrepresented areas in REUs.

4. What Should People Get Out of an REU

Any successful program must be rewarding for all involved. In particular, if we want to expand research experiences for undergraduates in the US, we must pay attention to the ways in which it can be rewarding for the *others* (i.e., other than the participants) involved in such programs. This will suggest how we should build new programs in different places.

Thus we begin with what mentors might get out of REUs. We consider this a central question. Our goal is to shift the culture of mathematics in the US, which requires convincing our colleagues. Faculty organizers should enjoy the experience, feel rewarded by their interactions with the students, and feel appreciated by our institutions.

Non-faculty mentors often get financial compensation (certainly important!), but not at a level commensurate with the time and effort they put in. Thus it is essential that they do not feel overworked, and feel supported by faculty organizers. This should be an important step in their careers — something that will contribute to their success in the future. Perhaps it will be useful in getting a faculty position, and in being successful in that position (in terms of research, teaching, *and* service). Perhaps they will work in the wider world, and take a leadership role working in groups — this experience should help them both get such jobs and to thrive in them.

Finally, we should ask what is in this for students. In this case, we wish to break the question up into two very different (and equally important) *subquestions*: what do students think they (want to) get out of it? And what do students *actually* get out of it? We should structure REUs so as to address both aspects.

Things students may want to get out of an REU:

- find out if I want to go to graduate school.
- find out if I want to go to graduate school in mathematics.
- do something productive with my summer.
- learn some interesting math.
- have fun.

One common motivation is: "My friends in other subjects are doing research, and I think I'm supposed to do it too. Perhaps I think that it will help me get into graduate school." Students often ask about opportunities for undergraduate research when choosing a college. They are often impressed by people who have a "publication". Some colleagues take this as a reason why undergraduate research is an unhealthy thing to encourage. But instead, we should use this motivation to deflect the students into getting something useful out of the experience!

There are many things we may want students to get out of an REU. For the authors (and SURIM), these are:

- a positive experience of engagement with mathematics.
- positive new relationships with mentors and peers.
- an experience of how doing mathematics is different from taking classes — both the joys and the frustrations.

Anything else is icing on the cake.

In particular, our goal is *not* to produce publications, and we would not measure the success of the programs by how many publications are produced. (Of course, the reader might accuse us of double standards, as we have emphasized the publication of a paper as one of the measures of success of one of our experimental models in the previous section! Again, we emphasize, this publication is not *the* measure of success of that group, it is merely *one* of them, and was certainly not planned for or even expected!)

We encourage the readers to write down what *they* think students should get out of their REU. *The answers should not be the same for all REUs!* The list should be concise. What is the *bare minimum* you would want to achieve in order to consider the summer a success for the student?

5. REUs and the Culture of Mathematics

The title of this section coincides with the title of one of the panels of the New Directions for Mathematics REUs conference in MHC in June 2013. It seemed appropriate after 25 years of REUs to reflect on the role of this institution on the *culture* of mathematics.

It is tricky to give a rigorous definition of this last notion. However, in this section we will attempt to jot together some seemingly random thoughts that might give an idea of what is meant by this idea. The unifying feature of this section is that the culture of mathematics in the US has shifted thanks to REUs.

The culture of mathematics includes the way in which mathematics is done. How have REUs affected this aspect? We believe that REUs train future mathematicians to be more collaborative and open to free exchange of ideas. This is a direction Mathematics is moving in, as more papers in pure mathematics are being coauthored than before. A typical REU is conducted in *teams*. Ideas are shared freely among the team members and the mentor, as well as amongst the different teams in the same institution. The success of the project therefore crucially hinges upon a deeply collaborative approach. It seems to us that (at least some) REU participants tend to carry this spirit with them forward into their future career. One should not underestimate this positive effect of REUs on our profession, especially in the internet age where cut-throat competition for priority has deeply undermined the more relaxed atmosphere in which mathematics was once conducted.

What is more, those areas of mathematics that are well-represented in REUs seem to have been visibly positively affected in the sense alluded to in the previous paragraph. For instance, a number of the young leaders in algebraic geometry have come from REUs, and the atmosphere in this field of mathematics has benefited accordingly. On the other hand, geometric analysis has seen very few REU programs over the past three decades, and is rather well-known for its particularly competitive atmosphere. Of course, such a statement is at best anecdotal, quite subjective and not quite scientific. But we mention it mainly in order to raise this possibly controversial issue.

Viewed in this light, a positive aspect about the new models presented in Sections 2.2–2.3 is that they have the prospect of involving many more people in the "REU world" and thus increase the positive effects on the culture of mathematics both inside and outside this world. This is so for two reasons. First, if indeed these models are applicable in R1 universities, this may increase the number of R1 faculty and graduate students involved in this kind of collaborative activity. Given the scarcity of REU sites in R1 universities (see Section 2.2), this also has the potential of dramatically increasing the number of REUs and thus the number of students affected. Second, by involving graduate students in R1 institutions, the prospects of having an influence on the profession are increased: those graduate students are precisely the faculty members of tomorrow.

Mathematics is arguably among the most social of all subjects. REUs precisely allow us to share and prove this to our students. Clearly, much mathematics is also done in isolation, and some individuals may not thrive in an REU environment. But the overall media stereotype of the "lonely genius" might, unfortunately, scare away from our profession individuals who might be extremely influential and important for the advancement of mathematics.

REUs also attract people that might otherwise not have entered our profession. For example, it gives an opportunity for students from less well-known institutions to strengthen their research background and credentials, and thus to have a shot at top grad schools. This should not be underestimated, as it influences the human fabric of our profession.

We can deliberately change the culture of mathematics. REUs have proven to have a positive effect on the culture of mathematics. Therefore, giving serious thought to ways in which REUs can evolve in the future is paramount.

Acknowledgments

The authors are grateful to the numerous referees for their very careful reading and comments that improved the exposition. They also thank O. Chodosh, V. Jain, J.P. May, J. Slimowitz Pearl, and S.

Wolpert for helpful comments. The authors wish to thank Jeremy Booher, Gunnar Carlsson, Otis Chodosh, Chris Henderson, Seungki Kim, Ryan Lewis, Sam Lichtenstein, Daniel Litt, Cary Malkiewich, Evita Nestoridi, Khoa Nguyen, Nancy Rodriguez, Niccolo' Ronchetti, and Simon Rubinstein-Salzedo without whose support and dedication SURIM would not have been possible over the past three years. The authors are also grateful to Stanford's Office of the VPUE and Stanford Department of Mathematics for funding SURIM, and to all the staff and faculty members at Stanford that helped make SURIM a success. Y.A.R. thanks Alex George for help in gathering data for Table 1. The NSF supported this research through grants DMS-1206284 (Y.A.R.), and DMS-1100771, DMS-1159156 (R.V.). Y.A.R. was also supported by a Sloan Research Fellowship, and R.V. was also supported by a Simons Fellowship.

References

[1] O. Chodosh, V. Jain, M. Lindsey, L. Panchev, Y.A. Rubinstein, On Discontinuity of Planar Optimal Transport Maps, *J. Topology Anal.*, **7** (2015) 239–260.

[2] D. Cristofaro-Gardiner, The Berkeley Summer Research Program for Undergraduates, talk at "New Directions for Mathematics REUs", Mt. Holyoke College, June 2013; "The Berkeley Summer Research Program for Undergraduates": One model for an undergraduate summer research program at doctoral-granting universities, this volume.

[3] J.F. Grcar, Topical Bias in Generalist Mathematics Journals, *Notices Amer. Math. Soc.* **57** (2010) 1421–1424.

[4] J.P. May, Fifteen Years of the REU and DRP at the University of Chicago, this volume.

[5] H. Miller, A Project Laboratory in Mathematics, talk at "New Directions for Mathematics REUs", Mt. Holyoke College, June 2013; K. Lin, H. Miller, A laboratory course in Mathematics, this volume.

[6] Carnegie Research I Universities (the 1994 list), https://math.la.asu.edu/~kuang/ResearchI.html (accessed in July 2014).

[7] NSF funded REU Sites in Mathematical Sciences, http://www.nsf.gov/crssprgm/reu/list_result.jsp?unitid=5044 (accessed July 2014).

Chapter 5

The Berkeley Summer Research Program for Undergraduates: One Model for an Undergraduate Summer Research Program at a Doctorate-Granting University

Daniel Cristofaro-Gardiner

Mathematics Department, Harvard University
Cambridge MA, USA
gardiner@math.harvard.edu

Historically, mathematics REUs have mainly been offered by institutions that specialize in undergraduate education. The purpose of this article is to offer one model for an undergraduate summer research program at a large research university by communicating and discussing the details of the UC Berkeley Topology, Geometry, and Operator Algebras Summer Research Program for Undergraduates. The author helped organize and design this program, and twice served as the research group mentor for the symplectic geometry group.

1. Preliminaries

1.1. The program and its funding

The "UC Berkeley Topology, Geometry, and Operator Algebras Summer Research Program for Undergraduates" was an 8-week program that ran at Berkeley in the summers of 2012 and 2013. In 2012, twelve undergraduates participated, and in 2013 the program was expanded to eighteen undergraduates. The program was geared

MSC 2010: 01A67, 01A80 (Primary)

towards rising juniors and seniors, and students were recruited nationally. Participating students were placed in research groups consisting of six undergraduates and a graduate student mentor; thus, in 2012 there were two graduate student mentors and in 2013 there were three. Graduate students were then advised by a faculty supervisor.

The program was funded by an NSF Research Training Group (RTG) grant. NSF RTG grants are meant to fund undergraduates, graduate students, and postdoctoral associates in research groups that are centered on a common interest [6]. These grants can be quite large (many are multi-million dollar grants), and they are not meant to be spent just on summer undergraduate research; a chart with the financial figures from Berkeley's grant is provided in Section 7. In 2009, an RTG in "Geometry, Topology, and Operator Algebras" was established at Berkeley with two faculty members as PIs and several more faculty as associated senior personnel. The grant funded many activities in addition to the 2012 and 2013 summer undergraduate research program. For example, it funded summer workshops for graduate students, a for-credit term-time undergraduate research seminar, and it provided summer research support for graduate students. The grant was renewed in 2014, and the summer research program for undergraduates was a key component of the proposal for renewal. We anticipate that the program will run again in the summer of 2015, and continue subject to funding.

1.2. Program designers and a disclaimer

The 2012 version of the program was designed by Michael Hutchings, Dan Pomerleano and the author. The designers of the 2013 program were Adam Boocher, Michael Hutchings, Dan Pomerleano, Pablo Solis, and the author. While we have aimed to coordinate with the other organizers, this article ultimately presents the author's point of view on the program.

1.3. Guide to the document

This article is organized as follows. In the next section, we highlight the unusual aspects of the Berkeley program. Then in Section 3, we discuss the details of the program in considerable depth. To illustrate these concepts, we present a specific research project in Section 4 that was successful. The remaining sections are devoted to reflecting on the program and analyzing it. In Section 5, we discuss the main challenges that came up while running the program and propose some potential solutions. Student feedback on the program is analyzed in Section 6, and issues involving the sustainability of the program are discussed in Section 7. We end in Section 8 by comparing the program to research programs at other doctorate-granting universities.

2. Unusual Aspects of the Program

We begin by highlighting some aspects of the program that to our knowledge are nonstandard, and we explain the philosophy behind these choices.

2.1. Program primarily run by graduate students

Graduate students occupied a central role in Berkeley's program. These students met with program participants every day and handled almost all organizational and advising duties.

Involving graduate students is a natural choice for a large research university that is interested in offering an undergraduate research program. Graduate students improve their own research skills by teaching others how to do research, and they gain useful experience in being research group leaders. Many graduate students are also interested in undergraduate education, so having the opportunity to participate in an undergraduate research program represents a valuable professional opportunity. Additionally, faculty at large research universities often use summers to travel to conferences and to focus on research, so an 8 or 10-week research program at a large university meeting almost every day would probably not be a good fit without substantial graduate student contributions.

2.2. "Underrepresented" topics

We offered research groups on "homological algebra" and "contact and symplectic geometry" in 2012, and "homogeneous spaces", "symplectic embeddings", and "computational commutative algebra" in 2013. Some of these topics belong to the "underrepresented" subfields of mathematics that were discussed at the "New Directions for Mathematics REUs" conference at Mount Holyoke in 2013 [7]. To briefly elaborate on what was discussed, there are important branches of mathematics that seem to be underrepresented among REU programs [7]. It was observed at the conference, e.g. in [2, 7, 8, 12], that REUs promote a spirit of collaboration and can shift the cultures of fields by seeding these fields with enthusiastic researchers who are excited about the free exchange of ideas. It is therefore highly desirable to offer REUs on a wide range of topics. It is our hope that our program will inspire more REUs in these underrepresented fields.

3. Program Specifics: A "User's Guide"

We now provide more specifics about how the program was run, and we also provide our reflections on these details. We go into considerable detail; ideally, this section can serve as a primer for readers interested in running a similar program.

3.1. General information about the positions to be filled

Our model called for one graduate student mentor for each six undergraduates, and one faculty advisor per graduate student. All current Berkeley graduate students were eligible to apply, even students with an expected graduation date before the start of the program (in fact, one of the mentors in 2012 and one of the mentors in 2013 received their PhD during the spring of the program year). In 2012, there were 12 undergraduates, and in 2013 there were 18.

The time commitment required of graduate students was substantial. Graduate students were asked to complete almost all the organizational duties required to set up the program, design research projects in conjunction with faculty and supervise them, meet with students essentially every day for 8 weeks, and organize and attend some social activities. We will spell out these duties in more detail in subsequent sections, but in general terms graduate students were asked to spend around 30–40 hours per week on the program during the 8 weeks when it was running and probably 20 hours in total setting up the program before the students arrived.

The time commitment required of the faculty supervisors was much more modest. By far the most important role faculty filled was to help the graduate students pick appropriate problems. Faculty also consulted on important decisions, and met with the students for one or two hours during the program. In general, the time commitment required of faculty was not meant to exceed 10 hours total across the summer.

Graduate students were paid around $10,000 for the summer, and faculty were asked to volunteer their time.

3.2. Organizational aspects of the program

3.2.1. *Recruiting graduate student mentors and faculty advisors*

Choosing the right graduate student mentors and the right faculty advisors was essential for the success of the program.

For selecting the graduate student mentors, one advantage was the large potential applicant pool. Berkeley has many graduate students, and most who were approached were interested in the program. Moreover, the author (who handled this selection process) knew many of these students personally, and conducted interviews. As such, it was not hard to select graduate students who seemed to have the potential to do an excellent job. For finding strong faculty advisors, the key was finding enthusiastic faculty who could make a significant contribution in a small amount of time. Once the graduate student supervisors were selected, it was left to them to find such faculty members. To succeed at this, being willing to ask many faculty was important. It was also important to communicate the goals of the program and explain how faculty could help very clearly. That said, finding a sufficient number of strong faculty advisors was definitely a challenge. For example, one of the graduate students asked two different faculty members for their suggestions and received only minor feedback.

3.2.2. *Designing the program website and recruiting undergraduates*

In both years, the first step to getting the program off the ground was to design the program website. While this was done almost entirely by the graduate student organizers, faculty input was also very valuable at this stage. To elaborate, the program website contained all the essential information about the program, including how to apply, and also contained loose descriptions of the various research projects. Faculty approved the website before it went live. It is worth emphasizing that designing the website can be somewhat subtle: indeed, in 2012 there was some confusion about the role of faculty in the program that came from unclear wording on the website.

Undergraduates were actively recruited nationally. Specifically, graduate students designed an electronic poster for the program, and emailed the mathematics departments at approximately 50 research universities and 50 primarily undergraduate institutions to advertise the program. The graduate students also arranged for the program to be listed on the AMS website. One weakness of our recruitment process was that we probably did not take enough steps to actively recruit underrepresented minorities and women. This is further discussed in Section 5.5.

3.2.3. *Selecting undergraduate participants*

Both years that the program ran there were approximately 10 undergraduate applicants for every spot. We aimed to have a diverse group of students, hailing from a wide variety of undergraduate institutions, who would interact well with each other and be able to handle the considerable demands of the program. We also wanted to be sure that at least a few students from Berkeley were included, in order to take advantage of potential synergy with other educational opportunities offered through the RTG grant.

Applicants were selected by the graduate students, with faculty occasionally giving input for difficult decisions. The application consisted of letters of recommendation, a transcript, and a personal statement. In certain close cases, a phone interview was also conducted. Each graduate student had considerable latitude in choosing his group, although efforts were made to coordinate the choices in order to ensure that the general goals for the selection process described in the previous paragraph were met.

During the selection process, no one aspect of the application was emphasized over any of the others. After the fact, it is interesting to reflect on which selection criteria seemed to have the most predictive value. Certainly the transcript is extremely important and should be read carefully. Here, the emphasis should be on ensuring that the student is adequately prepared; many students who had completed only a few upper division math courses struggled, while participants who had excellent grades in a wide variety of advanced courses rarely did. Very strong faculty recommendations also have excellent predictive value. The personal statement is probably of less use. The author also found the phone interview to be of limited help, essentially because all the interviewed applicants performed well. This was potentially because the interview questions were on the "soft" side, so a more rigorous phone interview could potentially be more useful.

3.2.4. *Arranging for room and board; student stipends*

The graduate student organizers also arranged for housing and meals.

For stipends, undergraduates were given the option of living in dorm-style housing, with board covered, and receiving a $1,000 stipend, or receiving a $3,000 stipend which could be spent towards housing they selected. This second option was primarily designed for Berkeley undergraduates participating in the program, since many upperclassmen at Berkeley already live off campus during the academic year. Most students selected the $3,000 stipend.

3.2.5. *Social activities*

The graduate students also organized social activities for the students during the program. These happened frequently, and included activities around Berkeley, such as group runs, ice cream outings, dinners, frisbee, and hikes, and occasionally more elaborate activities like group trips. In 2013, a joint hike was organized with Stanford's Summer Undergraduate Research (SURIM) program. Part of the purpose of this hike was to give the program participants the chance to get to know the Stanford students in advance of a joint conference, discussed further in Section 3.5.1.

3.3. Mathematical aspects of the program

3.3.1. *Choosing research projects*

Graduate students designed the research projects in conjunction with their faculty supervisors. We found that choosing the right research projects was critical for the success of the program, so we now elaborate on how this was done. First, graduate student supervisors were asked to come up with several potential projects suitable for undergraduates and then ask their faculty supervisor for their input. These projects were expected to center around original research; expository projects were not acceptable. Faculty supervisors were also encouraged to suggest any other projects that came to mind. From these discussions, a list of potential research projects was created for each group; although there were six undergraduates per research group, students were expected to work together to some extent, so each graduate student supervisor had between two and around six potential projects.

It should be clear from above that a significant share of the responsibility for choosing good research projects fell on the graduate students. Moreover, in general graduate student supervisors received a fairly minimal amount of structured help from faculty, e.g., there was no formal training on how to select good projects for undergraduates. Thus, in making sure that good research projects were selected, it was hugely important to select the right graduate student mentors and faculty advisors. We refer the reader to Section 3.2.1 for advice on how to handle this selection process. To give an example of good faculty advising, the author sent his advisor a list of 10 or so potential projects, and his advisor was quite firm in pointing out projects that were not feasible, flagging ideas that seemed good, commenting on subtleties the author had not considered, suggesting improvements, and proposing other projects. This was all done via email, and so was handled in a time efficient manner.

The author also found that it was possible to suggest projects that complemented his own work well. For example, some of the computations he assigned to a student in 2012 inspired him and his collaborators to work on a project which eventually appeared in the *Journal of Topology* [3]. This is further discussed in Section 6.3, and more advice on choosing appropriate research projects is given elsewhere in this article. For example, we provide details of a project that went well in Section 4, and we discuss specific strategies for designing good research projects in Section 5.1.

3.3.2. *Lectures, problem sets, and problem sessions*

The program had a substantial lecture component, with each graduate student giving approximately 20 hours of lecture. This was done in part because of the program's emphasis on "underrepresented" REU topics. One reason that topics like differential geometry are not usually the focus of REUs is because they are perceived to have a high barrier to entry. The idea behind the lecture component was to ease students' transitions into doing research in what are normally graduate level topics. The lecture component was also meant to enrich the students' overall experience by giving them appropriate context for their research. The organizers also hoped that the research projects and the lectures would be synergistic, in the sense that students' research projects would cause them to be excited about their field, which would in turn cause them to be more enthusiastic in understanding the details of the lectures.

In 2012, the program consisted of 4 weeks of lecture (meeting every day) followed by 4 weeks of research. After reviewing student evaluations, we found that this format was not ideal: students were dissatisfied with the amount of time it left for research, and students found it hard to stay motivated to understand the lectures for 4 weeks. Because of this, in 2013 we switched to a more integrated format, where students started their research essentially on day one and lectures were interspersed throughout. While the total amount of lecture time remained roughly the same, this format seemed to be much more effective.

In the 2012 version of the program, there were weekly problem sets and frequent problem sessions. While these seemed to be somewhat effective, we decided to significantly scale back this aspect of the program for 2013 because of feedback from the instructors and the students from 2012.

3.3.3. *Supervising the research projects*

Graduate students were also in charge of supervising research projects.

The undergraduates were asked to work in groups. Some care was required in arranging for appropriate groups, and this was mainly left to the graduate students' discretion. Graduate students met with students to discuss research for between 40 and 80 hours over the course of the program, and generally tried to have at least one research meeting each day. The precise nature of the graduate student supervision was left to each mentor's discretion.

It seems that choosing to meet every day, and for many hours each day, is a somewhat controversial choice (we will briefly compare the Berkeley program with similar programs in Section 8). For one, as discussed in Section 7, this format puts considerable strain on the graduate students. Moreover, it is far from clear that this is the best arrangement for the undergraduates themselves, since learning to work independently is part of learning how to become a good researcher. In the author's experience, meeting every day is probably not essential, especially in the latter weeks of the program (front loading the supervision of the undergraduates seems also to be common in similar programs, see Section 7). Nevertheless, a significant amount of face time between the undergraduate researchers and their graduate student supervisors is important. Eight weeks is simply not much time, especially for a program like Berkeley's that aims to have high quality writeups containing original research on graduate level topics. Thus, in running a program like Berkeley's, the author would suggest meeting around four times per week in the first few weeks of the program, and then at least three times per week the rest of the time.

Supervising the research projects effectively was one of the most difficult parts of the program, and some of the challenges that came up as well as potential solutions are discussed further in Section 5.1. In broad terms, it was critical to make sure that research projects were manageable and program participants stayed focused on completing them.

3.3.4. *Writeups*

Student participants were asked to write up their work. In 2012, this requirement was not pursued very aggressively, and some students' writeups ended up being weak. As a result, in 2013 this was made an emphasis of the program. This is discussed further in Section 5.2.

The writeups were expected to present original research and be of publishable quality. To give more specifics, the 2013 symplectic group produced three writeups. These were between 7 and 30 pages, and were written in the format of research articles; two were submitted for publication, and we expect that the third will be as well. One [9] has been accepted for publication in the *Ramanujan Journal*, and another [5] was recommended for *Involve*. For reference, *Involve* is a journal which publishes high quality articles by undergraduates (more senior coauthors are welcome too), while the *Ramanujan Journal* publishes original articles in fields that were influenced by the mathematician Srinivasa Ramanujan. The computational commutative algebra group also produced a paper [1] with their graduate student supervisor that has been recommended for publication in the *Annals of Combinatorics*, a top combinatorics journal.

We did not provide the students with any formal training on how to write a research article, but the graduate student mentors worked closely with their undergraduate advisees in helping them polish the writeups and make them suitable for public dissemination. It was important to supervise much of this writeup process and start it early. For example, in 2013, we asked the students to begin writing up their work in the seventh week of the program. It proved infeasible to complete the writeups within the 8-week time frame of the program, so graduate students stayed in touch with their undergraduates after the program to put the finishing touches on this work. In some cases, these ongoing conversations lasted up to a year after the program (in fact, some are still in progress).

3.4. Program blog

In the 2013 program, we also ran an online blog [10] for several weeks. Pablo Solis (the graduate student mentor for the "homogenous spaces" group) made the most heavy use of the blog; for example, Pablo posted expository notes, and had several other students contribute to the blog in the early stages of the program. He found that it was a good way to evaluate to what degree students were digesting the material. Pablo also felt like if the program were run again, the blog should have been used more. For example, having students blog on what they are learning would be a great way to ease them into the writing process.

3.5. Presentations

3.5.1. *Student presentations and a joint conference*

The program participants were asked to give presentations. This was particularly emphasized for the 2013 program. At the end of the 2013 program, there was a joint conference with the SURIM program at Stanford where each research group gave a 20-minute presentation. To train for this, the Berkeley students gave a practice presentation on their research to the other program participants. The program participants were also asked to give expository talks about their research field.

Students received feedback from their peers and from their research group mentors about how to improve their presentations.

3.5.2. *Colloquium series*

In 2013, the program also had a weekly colloquium series. This mainly featured mathematical talks by faculty members, but also occasionally addressed more practical issues: for example, the graduate student mentors ran a panel discussion on how to apply for graduate school and the NSF graduate fellowship.

3.6. Links with other programs

As mentioned in Section 3.2.5 and Section 3.5.1, in 2013 a hike and a conference were scheduled in conjunction with Stanford's "SURIM" program. These activities were highly successful, and it would be great to forge similar joint ventures in future versions of the program. For example, the program was in no way affiliated with MSRI's "UP" program, but in future years it would be desirable to forge some connections with this program, and to strengthen connections with SURIM.

3.7. Survey data

At the end of each summer, we asked students 10 specific questions about their experience. This was done anonymously via an online program, with a 100% response rate each time. We asked a combination of multiple choice and free response questions. The questions covered many different aspects of the program. For example, there were questions about the social activities, the timing of the research projects, and the quality of the graduate student supervision. We found this data quite useful, especially in making improvements between 2012 and 2013.

4. A Case Study

We now present a research project from 2013 that was supervised by the author in order to further illustrate the above program concepts.

In 2013, three students in the symplectic geometry group worked on a research problem involving the symplectic geometry of four-dimensional "toric domains". These are open subsets of $\mathbb{R}^4$ with a high degree of symmetry, and they have a natural symplectic form given by restricting the "standard" symplectic form on $\mathbb{R}^4$. It is interesting to ask when one four-dimensional toric domain symplectically embeds into another, and there is a new family of symplectic embedding obstructions, called embedded contact homology (ECH) capacities, that can be used to help answer this question. ECH capacities can be computed purely combinatorially for some toric domains in terms of lattice point counts in certain planar regions, making them particularly well-suited for exploration by undergraduates.

The three undergraduates first extended this combinatorial formula to more general toric domains by explicitly constructing certain symplectic embeddings. They then studied symplectic embeddings out of a family of toric domains from this class, and showed that ECH gives a sharp obstruction to embedding elements of this family into balls in some interesting cases, thus obtaining new results about symplectic embeddings. The students wrote up their work [5], and it was recommended for publication in *Involve.*

The project was specifically designed so that the students could work with many explicit examples without having to understand many of the intricacies behind the ECH capacities, which involve complicated ideas like Floer homology and pseudoholomorphic curves. But after completing the project, one of the undergraduates was motivated to learn more. He gave an expository talk on a related idea called "cylindrical contact homology" to his peers, and this student was so interested in learning about ECH that he asked the author to serve as a co-advisor for a senior thesis project on ECH, even though he was not a Berkeley undergraduate. The author agreed to do this. The undergraduate then decided he was interested in learning more about differential geometry and low-dimensional topology, and so applied for external fellowships to complete Part III at Cambridge. He won a Marshall scholarship, and will try to learn more about these topics during the 2014–15 academic year.

One of the other students in this group was motivated by the project to explore a classical tool for lattice point enumeration called the Ehrhart polynomial, which is also connected to ECH capacities of certain toric domains. The student proved several interesting results about this tool which were accepted for publication [9].

5. Challenges and Potential Solutions

We now discuss in more depth some challenges that came up while running the program and we mention how we attempted to address them.

5.1. Difficult research projects

Many of the program's research topics involved fields that are generally not studied until graduate school. As such, finding projects at a suitable level for students could sometimes be difficult. For example, some of the students in the 2012 program mentioned on evaluations that they had a hard time understanding the context for their research problems and sometimes found it demoralizing to work with tools that they had not yet fully absorbed.

We worked hard in 2013 to try to make it so that these issues would not come up. One way we attempted to do this was by making sure to choose "hands-on" research problems, as in Section 4. In general, problems were chosen for which one could compute many examples and make use of computer exploration, and efforts were made to simplify the technical demands of problems as much as possible. For example, the symplectic geometry group worked exclusively with open submanifolds of $\mathbb{R}^{2n}$, and many of their projects involved exploring certain symplectic embedding obstructions that were very combinatorial in nature.

We also tried to make the research problems more "dynamic". Specifically, we tried to have students start the research projects by completing some simple computations and solving relatively easy problems. Then, if they finished this quickly, we would give them something more challenging to think about. Similarly, if students seemed to be excessively struggling with their research problem, we would redesign the goals of the project to make them more attainable. In other words, we asked the graduate student mentors to play an active role in making sure that students were appropriately challenged throughout the program. Exactly how this was carried out was left to the discretion of the graduate student supervisor.

In the author's 2013 symplectic group, there was no one standardized way in which this process was handled. One of the groups, which consisted of three advanced students, was given as an ultimate goal a quite general conjecture; they were quickly able to prove it in many special cases, but the general case eluded them. Another group was given a specific problem at the beginning of the program, and they were able to complete it. This took the group essentially the full 8 weeks, and so the timing for the project worked out quite well.

In general, we found that it could be frustrating for students if they did not achieve their original goal; this came up for example with the first group from the previous paragraph. In the author's experience, if it is made clear to the students that this is a natural part of research, then students do not get too demoralized. The key is for the graduate student supervisor to make sure that at least some progress is made, and provide positive feedback when this happens.

5.2. Lagging enthusiasm for the writeup

As mentioned in Section 3.3.4, it was sometimes challenging to get students to produce high quality writeups. Writing a good research article takes time, and when many students imagine an ideal summer doing math, they do not think of spending considerable time writing up their work clearly and correctly. Moreover, some students can become disappointed if their projects do not pan out in full generality, and this can drain enthusiasm for a long writeup process.

We found that the students were in general highly motivated to see their work appear in print, but that it was also necessary to make it very clear to them that their writeups were mandatory, and would be read by the graduate student supervisors and ideally by Berkeley faculty as well. In the author's opinion, some direct supervision and lots of persistence were also important. The writeup phase for two of the three projects from the author's 2013 symplectic group lasted well past the end of the program, and it seemed likely that at least one of the groups would not have finished their writeup if they were not occasionally reminded by the author. Positive encouragement, especially for groups that were somewhat disappointed in their work, was also quite valuable.

Getting good writeups from the students also required choosing good projects, as in Section 5.1. We tried to ensure that every group completed some piece of original research, and this was done by making sure that projects were well chosen and well supervised.

5.3. Graduate student inexperience

We found that graduate student inexperience could sometimes be a problem. The graduate students who participated in the program had little to no experience mentoring undergraduates, and were generally close to the beginning of their own research careers. This sometimes made it difficult to choose good problems for their students to work on. For example, in retrospect one of the groups in the 2012 program worked on a problem that was simply too difficult. It was also sometimes challenging for graduate students to exude confidence that their advisees' research projects would progress well, since they themselves were new to research; we found that this was sometimes demoralizing for the undergraduates. Similarly, some of the undergraduates reported that it was frustrating to not have much faculty supervision.

We tried to address these problems in advance of the 2013 program. In general, we tried to make sure that faculty were involved in helping to select appropriate research problems, and we also tried to choose graduate students who were almost at the end of their PhD and who had successfully produced math research before.

5.4. Difficult students

In 2012, there was one student who did not always do what was asked of him throughout the program, and this came to a head at the end of the program when he refused to promptly submit a writeup and a faculty member had to get involved. In retrospect, the graduate student supervising the student should have contacted his faculty advisor for help much sooner.

5.5. Recruiting women and underrepresented minorities

As mentioned in Section 3.2.3, we aimed to enroll a diverse group of students. While we were successful in finding students from a range of educational institutions, we had less success finding applications from underrepresented minorities, at least anecdotally (unfortunately we did not collect data about this). One obvious way to address this is by stepping up attempts to actively recruit these students, see the end of Section 3.2.2. A potentially related problem was the low stipend in comparison with similar programs. We have no data to confirm this (and indeed, no students even contacted us about this), but it is certainly conceivable that low-income minority students were deterred from applying to our program because of a need for greater financial support during the summer. Perhaps the stipend should be increased, or a need based supplement made available.

We also would have liked to receive more applications from women: for example, in 2013 only about 1 in 5 applicants were female. While we were happier with the gender balance in the pool of students who ended up participating in the 2013 program (6 out of 18), this could be improved as well. For this, better recruitment as in Section 3.2.2 should help.

5.6. Excessive demands on graduate students

Another concern is that the program places excessive demands on the graduate students. Probably only advanced graduate students are generally capable of handling all the responsibilities of the program, and for such students time is often at a premium. The amount of time asked of the graduate student supervisors is definitely an issue and is discussed elsewhere in this article, see Section 7. We note, though, that all four of the graduate students who participated in the program successfully applied for postdocs after participating in the program, and accepted multi-year positions; these were at the University of Utah, Harvard University, Imperial College London, and the California Institute of Technology. See also Section 6.3 for more graduate student feedback.

6. Feedback

We surveyed the students after each program, and we now present some of the results.

In general, feedback was very positive. In 2013, all 18 undergraduate participants and all four graduate student mentors described the program as good or great on anonymous surveys (14 out of the 18 undergraduates and 3 out of the 4 mentors described it as great). We now discuss the feedback in greater depth.

6.1. Student feedback

The first set of results we discuss involve changes made to the program between 2012 and 2013. Specifically, after 2012, we reduced the amount of lecture and increased the amount of time devoted to research. As the charts at the end of this section show, these changes were effective. It is also interesting to note that the balance we struck between lecture and research was good, with 56% of students describing both the amount of lecture and the amount of research as "just right".

Most of the written feedback was positive. One student in the 2012 program wrote: "I really enjoyed my research project. At first it seemed too difficult, but after the lecture portion of the program, my project turned out to be at a good level for me." Another from 2012 said "I thought it was a great experience, and I learned more math than I thought would be possible in a single summer." A student from the 2013 program wrote "I will never forget this summer: the people I've met and the insight I've gained into math research in general. I had a wonderful time both socially and academically." A different student wrote that the program "was a great experience and has encouraged me to push myself more in math during the school year."

The suggestions for improvement were varied. One student in the 2013 program commented on our efforts to achieve a gender balance: "the fact that the split [between males and females] was so exact in each group seemed a bit artificial... and made me feel a bit uncomfortable." Another suggestion involved how the students' collaborations were managed: "I wish that the project could have been broken apart more such that there weren't three people working on the exact same problem at a time." There also seemed to be some concern that lectures were too advanced and moved too quickly. In a different direction, a student from 2012 suggested that their project "could have been more open ended".

Comparing student satisfaction with the amount of lecture. Question: "How did you feel about the amount of time devoted to lecture?"

Student Satisfaction: Lecture

DESCRIPTION	2012	2013
Too much	7	5
Just right	4	10
Too little	1	3

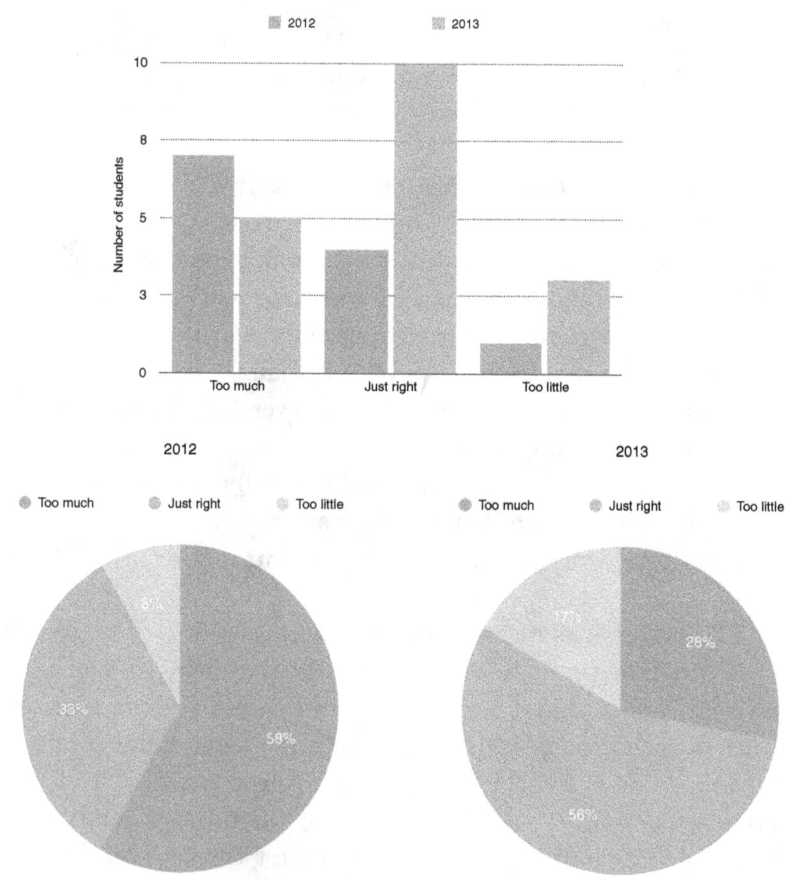

Comparing student satisfaction with the amount of **time devoted to research. Question: "How did you feel** about the amount of program time devoted to research?"

Student Satisfaction: Research

DESCRIPTION	2012	2013
Too much	1	1
Just right	3	10
Too little	8	7

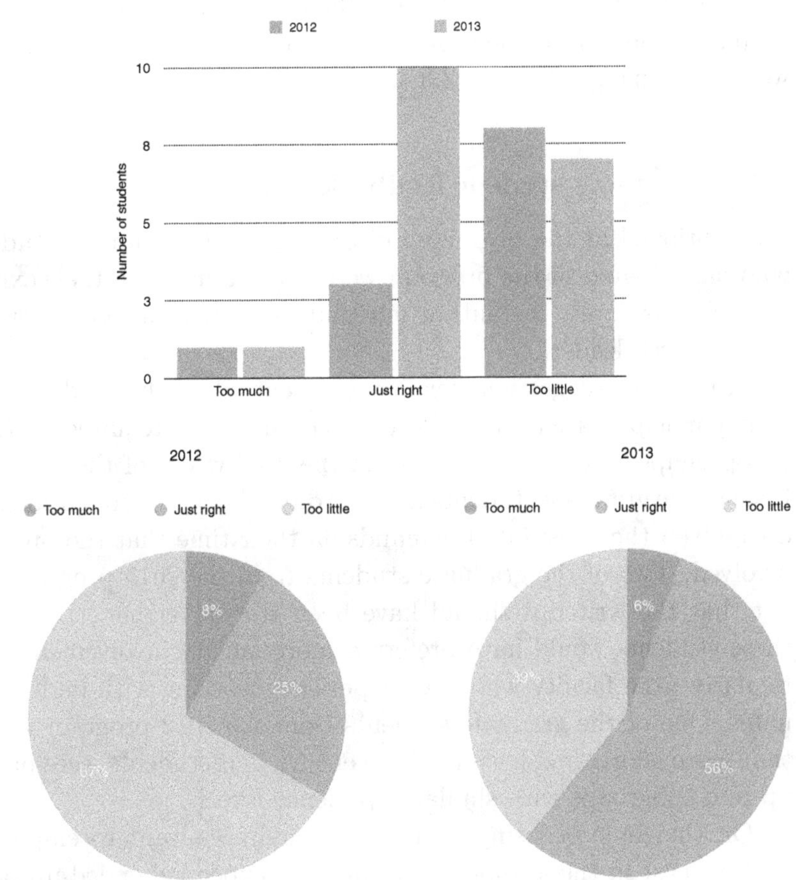

6.2. Faculty feedback

We asked Michael Hutchings, who was a faculty supervisor for the program in 2012 and in 2013, to comment on his experiences with the program. In general, he was happy with the program. When asked, he said that the time commitment requested of him was small, and supervising the program did not meaningfully affect the time he had for research. One of his graduate students participated in the program, and he did not think that the program affected the graduate student's ability to graduate on time. Hutchings added that he felt the most critical component for the success of the program was selecting the right problems. When asked what could have gone better, he mentioned the problems with the writeups from 2012 that were mentioned in Section 3.3.4.

6.3. Graduate student feedback

As mentioned at the beginning of Section 6, the graduate students who participated in the program were very happy with their experience. For example, the author felt that he grew as a researcher and as a research leader.

In terms of suggestions for improvement, all of the graduate student participants felt that the time commitment required of them was perhaps excessive, especially in the final weeks of the program. Each of the graduate students found it difficult to get enough research done given the substantial demands on their time that the program involved. Two of the graduate students from the 2013 program also felt that the writeups should have been started sooner, and one of these students would have preferred more faculty involvement, particularly from faculty who have experience working with undergraduates. One of the graduate students from the 2012 program wished that more efforts had been made to ensure that students were broken up into subgroups with similar experience levels.

On the topic of losing time for research, we want to emphasize that at least in the author's case the time expended ended up being extremely fruitful for his research program. To elaborate, the author's thesis work was primarily on topics not appropriate for an 8-week undergraduate research program. Thus, the author was motivated to learn about symplectic embedding problems in order to serve as an effective mentor. This has proved very fruitful. For example, as mentioned at the end of Section 3.3.1, the author wrote a paper about this topic that appeared in a top journal [3]; he also has completed several other projects related to this work that he expects will be accepted for publication, and he is currently working on a collaborative project with one of the top researchers in this field. None of this would have happened were it not for his experiences in the program, and his discussions with his students in the program have proved very fruitful for him. As mentioned in Section 3.3.4, another graduate student has had a strong paper [1] recommended for publication that he wrote with his students from the program.

7. Sustainability

While the program was designed with sustainability in mind (for example, we tried to arrange it so that the demand on faculty's time would be minimal), we anticipate two obstructions to the future success of the program. One is finding consistent funding. The program represented a relatively minor slice of the total RTG grant in the years that it ran, as Table 1 illustrates.

In this sense, then, the program is quite sustainable as long as a university can get such a large grant. There are several ways the program could be run with a smaller grant. The most obvious way to scale back spending would be to decrease the number of undergraduate participants. The graduate student salary worked out to around $30 per hour, and could also be slightly reduced.

Another obstruction to sustainability comes from finding graduate students who are excited about the program and willing to spend a lot of time. As mentioned in Section 3.1, the amount of supervision given

Table 1. Amount of RTG money spent on the program in the years the program ran (figures approximate). Note that the RTG grant also awarded $470,000.00 annually from 2009–2011.

	2012	2013
Total RTG money awarded to Berkeley	$470,000.00	$470,000.00
Money spent on the program	$60,000.00	$100,000.00

to the undergraduates was probably excessive and should be cut back (this would also make the program slightly less expensive). Also, it would perhaps make more sense to hire more graduate student mentors and reduce the time commitment for each mentor.

8. Comparison with Other Programs at Research Universities

As mentioned in the introduction, it seems that most undergraduate research programs are run at institutes that primarily focus on undergraduate education. We now briefly mention some other undergraduate research programs at large research universities. This is by no means supposed to be a comprehensive survey (we apologize in advance for any omissions), but it hopefully places the Berkeley RTG in an appropriate context.

Since 2000, the University of Chicago has been running an REU through its *VIGRE* program. The Chicago REU has grown over the years, and is much larger than ours. For example, it had 98 participants in 2009 [4]. While graduate students are also important to the Chicago model, faculty play a much larger role than in our model. The Chicago program also has students participate in an outreach program, for example by teaching high school students. Unlike our program, the Chicago program is only for University of Chicago undergraduates. Another difference is that while students in the Chicago program are also required to write a paper, their paper can be expository if they choose. The amount of required supervision in the Chicago program is also less.

The *Stanford Undergraduate Research Institute in Mathematics* (SURIM) program has been mentioned elsewhere in this article. This is a 10-week program for Stanford students that has been running since 2012. We benefited greatly from learning about this program from presentations by Yanir Rubinstein and Ravi Vakil at the "New Directions for Mathematics REUs" conference at Mount Holyoke [7, 12]. In many respects, SURIM is quite similar to the Berkeley program; one difference is that students have the option of working one-on-one with faculty.

9. Acknowledgments

It is a pleasure to thank Adam Boocher, Michael Hutchings, Daniel Pomerleano, and Pablo Solis for all of their help in designing and running the program, and for their suggestions for improving this article. Many of the ideas presented here came from discussions with them. We also thank the organizers of the "New Directions for Mathematics REUs" conference for inviting us to speak, and we thank the editors of this book for their encouragement. It is also a pleasure to thank Mark Peterson, Yanir Rubinstein, and the anonymous referees for their extremely helpful feedback on earlier drafts of this work.

This article was written while the author was a member at the Institute for Advanced Study. We thank the Institute for their generous support. Part of this support came from NSF grant DMS-1128155. Any opinions, findings and conclusions or recommendations expressed in this material are those of the author and do not necessarily reflect the views of the National Science Foundation.

References

[1] A. Boocher, B. Brown, T. Duff, L. Lyman, T. Murayama, A. Nesky, K. Schaefer, Robust Graph Ideals, to appear in *Ann. Combin.*

[2] C. Castillo-Chavez, The Power of Student Driven REU Programs: An Alternative Model of Learning Through Research, *Presentation at "New Directions for Mathematics REUs"*, Mt. Holyoke College, June 2013.

[3] K. Choi, D. Cristofaro-Gardiner, D. Frenkel, M. Hutchings, V. Ramos, Symplectic Embeddings into Four-Dimensional Concave Toric Domains, to appear in *J. Topology.*

[4] The University of Chicago Mathematics Department, VIGRE at the University of Chicago, http://www.math.uchicago.edu/~may/VIGRE/.

[5] M. Landry, M. McMillan, E. Tsukerman, On Symplectic Embeddings of Toric Domains, to appear in *Involve.*

[6] National Science Foundation, Research Training Groups in the Mathematical Sciences, http://www.nsf.gov/pubs/2014/nsf14585/nsf14585.pdf.

[7] Y. A. Rubinstein, Underrepresented Areas in REUs, *Presentation at "New Directions for Mathematics REUs"*, Mt. Holyoke College, June 2013.

[8] D. O'Shea, Types of Research Suitable for REUs, *Presentation at "New Directions for Mathematics REUs"*, Mt. Holyoke College, June 2013.
[9] E. Tsukerman, Fourier-Dedekind Sums and an Extension of Rademacher Reciprocity, to appear in *Ramanujan J.*
[10] P. Solis *et al.*, Berkeley Summer REU 2013, http://berkeleyreu2013.blogspot.com.
[11] R. Vakil, Experimenting with New Ideas for REUs, *Presentation at "New Directions for Mathematics REUs"*, Mt. Holyoke College, June 2013.
[12] R. Vakil, The Spirit of Collaboration and Free Exchange of Ideas in the Mathematics Community, Seen Through REUs, *Presentation at "New Directions for Mathematics REUs"*, Mt. Holyoke College, June 2013.

Chapter 6

Fifteen Years of the REU and DRP at the University of Chicago

J. Peter May

Mathematics has become one of the most popular majors at the University of Chicago, now accounting for 10% of all undergraduates. This article describes how that came about, with focus on relevant new programs. A remarkable atmosphere of genuine excitement about mathematics has developed in step with the expansion of these programs. Perhaps only some of the details can be emulated elsewhere, but the crucial idea of collaborating with participants in developing attractive programs can be emulated anywhere.

1. Background and Statistics

This article can be viewed as a report on an NSF sponsored experiment.[1] The University of Chicago had two consecutive VIGRE grants, on which I was PI, starting in June of 2000 and continuing until September of 2011. Those grants funded the first twelve years of our REU and several smaller new programs. With considerable internal University of Chicago support, all of the programs initiated with the help of those grants have continued operation the past three years. With the help of a new RTG grant, all will continue at least through 2019. All except the REU appear to be sustainable indefinitely, without external support.

With a focus on undergraduates, we will try to answer what is perhaps the most important question first. What clear, definite,

MSC 2010: 01A65, 01A72 (Primary)

[1] It is also an update of a similar article from eight years ago [2].

quantifiable changes can reasonably be ascribed to the new NSF sponsored programs at Chicago? Only then will we describe what those programs are, how they operate, and how they might be emulated elsewhere. For the last, we will focus on a relatively inexpensive program, the Directed Reading Program (DRP), that can have significant impact.

1.1. Statistics

Chicago has a small liberal arts college, with around 5000 students, embedded in a large research university. Undergraduates have over 100 majors to choose from. The immediate statistical question asks what number and percentage of those students major in mathematics. Here is the answer over time, starting in 2001. That is the last year before the new programs began to have significant impact, and it is also the first year for which we have collected full statistics.[2]

Number and percentage of mathematics majors among graduates, 2001–2014:

	2001	2002	2003	2004	2005	2006	2007
Total	988	947	1,000	1,013	1,072	1,094	1,158
Math	49	57	77	75	76	71	90
Percent	5.0	6.0	7.7	7.4	7.1	6.5	7.8

	2008	2009	2010	2011	2012	2013	2014
Total	1,185	1,207	1,209	1,271	1,242	1,304	1,385
Math	77	96	87	105	103	129	137
Percent	6.5	8.0	7.2	8.3	8.3	9.9	9.9

[2]Roughly 5% of Chicago's undergraduates have been mathematics majors since I arrived in 1967, but time constraints have prevented me from trying to obtain precise earlier data.

Visibly, there was a 40% increase in the total undergraduate population over these years and a 180% increase in the total number of mathematics majors. The expansion of the college was planned. Despite the expansion, it appears to us that the quality of undergraduate students has improved significantly over this period. Test scores are corroborative.

It is reasonable to ask how mathematics compares in popularity with other subjects. Here are the top five majors in terms of numbers of bachelor degrees at Chicago over the past eight years:

Subject	2007	2008	2009	2010	2011	2012	2013	2014
Economics	234	211	257	260	298	262	270	320
Biological Sciences	138	134	153	128	170	153	163	203
Political Science	109	111	108	137	148	131	155	140
MATHEMATICS	90	77	96	87	105	103	129	137
English	89	77	88	81	80	92	87	88

One might then ask if this trend is sustainable. The spring count of mathematics majors first passed the 200 mark in 2006, when it reached 227, and it only passed the 300 mark in 2011, when it reached 315; it already passed the 400 mark in 2014, reaching 411. Thus, rather alarmingly in terms of our capacity to deal with the numbers, we are on track for still further increases in the number and percentage of mathematics majors in the next few years.

Among US schools with the wealth of options for majors available at Chicago, surely none has percentages and numbers like that. One top university about the same size as Chicago has rightly celebrated an increase to a total of around 40 mathematics majors per year. That is real progress, but it is not on the same scale. Online data and data from graduate admissions[3] show that only MIT has a comparable total number of mathematics majors.

[3] As cochair of graduate admissions, I have long kept track of the numbers of applicants to Chicago from all other schools.

One might then ask about the quality of undergraduate math majors and, specifically, about how many are going on to graduate study at top universities. Virtually all of the top US mathematics departments have current graduate students from Chicago, the vast majority of them having participated in the REU or the DRP. Around 80 students graduating in the past six years, 2009–2014, went on to graduate study in mathematics, attending nearly 40 schools. The most frequent choices were Michigan (13), Berkeley (8), Harvard (4), MIT (4), Stanford (4), Texas (4), Maryland (3), UCLA (3), and Yale (3). Two each went to CalTech, Cambridge, Columbia, Princeton, UIC, and UIUC.

Of these 80, 15 were women. The percentage of women is not nearly as high as we would like, but the number is significant. It includes two winners of the Alice T. Schafer Prize (2010 and 2014). Two other winners (2008 and 2013) were or are Chicago mathematics graduate students.

On average, over 10% of Chicago mathematics majors go on to graduate study in mathematics each year, and over 10% go on to graduate study in other sciences. That only counts those we know are going on to graduate study immediately. A retrospective study of REU participants through 2009 showed that, while fewer than 50% stated they were going on to graduate study at the time they graduated, nearly 80% ended up doing so eventually.

1.2. The undergraduate program

What changes accounts for these statistics? Let's start with what did *not* change.

Were there substantial changes to the mathematics curriculum? For better or for worse, the answer is no. Under the pressure of undergraduate demand, courses and tracks have been added, but the basic curriculum has not seen any significant changes, certainly nothing that might seriously affect the statistics. It is worth emphasizing that all undergraduate mathematics courses at Chicago are taught rigorously, with proofs. Some changes made elsewhere were not made at Chicago.

Were there substantial changes in the way mathematics is taught? There has been one significant change, the introduction of inquiry based learning, or IBL, classes (the Moore method). These have definitely had an effect. They are primarily freshman honors calculus classes, which do not provide the fast track for the most advanced incoming students, and most of the honors calculus classes are still taught in the traditional lecture format. The IBL classes do not affect enough students to account for the statistics, but they have had a significant synergistic impact. There have been no other significant changes. In particular, again for better or for worse, Chicago has been backward with respect to technological innovation in teaching.

Have there been changes of requirements that might have raised enrollment in mathematics classes and might have helped attract mathematics majors? Quite the contrary. There have been some changes in requirements, but these changes have lessened rather than increased the requirements in mathematics. Another change that might well have decreased the numbers is the institution of a new Computer Science major in 2006; there is also a Statistics major.

1.3. New programs and changes in the culture

We will focus on two programs with direct impact, the DRP (directed reading program) and the REU. Two other seemingly unrelated programs, the VCA (Vigre Course Assistant) for undergraduates working in our teaching program and the warm-up program for incoming graduate students (WOMP), have an indirect synergistic effect.

This article will have a theme. The key to the increased popularity of mathematics at Chicago is something that is not quantifiable, namely the development of an extraordinary rapport among undergraduates, graduate students, and even postdocs over the past 15 years. Before 2000, these groups were socially separate. Now they are not. Social activities bringing undergraduate and graduate students together are sprinkled throughout the year. For example, just last Saturday (October 25, 2014), there was a “‘math study break” for undergraduates organized by graduate students in our AWM chapter. It was advertised with the enticement that “Math grads will be on hand to help you ~~cheat~~ (!) study for the mid-terms”. For another example, there is an annual and nowadays quite elaborate beer skit (make fun of the faculty — you can see some on the web). It is put on in late spring by second year graduate students. Before 2000, undergraduates had nothing to do with it and rarely if ever attended. Now it is a tradition that undergraduates serve a barbecue before the skits and attend it in droves. We have no room big enough to hold the resulting crowd.

Social intermingling of undergraduate and graduate students also takes place in the quarterly evening talks of the DRP and the occasional lunches and dinners with talks that are a regular part of the REU. Such socializing is intertwined with the mentoring of undergraduates by graduate students that is at the heart of the DRP and REU. Neither program could exist without the melding of levels, vertical integration, to use the slogan, that pervades the mathematics program at Chicago. This plays a central role in both the attractiveness of mathematics to Chicago undergraduates and the success of the graduate program.

A related theme is that our programs are bottom-up rather than top-down. They are driven by student demand, and student demand is driven by the cultural change. In particular, faculty do not recruit undergraduates to our programs. Past participants, graduate students, and postdocs do that job. It is all word of mouth.

2. The DRP and VCA Programs

2.1. Background of the DRP; the VCA

We will start with the DRP because it is inexpensive and implementable anywhere, or at least anywhere that has a sizable pool of graduate students. In fact, there are current DRP programs that are modeled on Chicago's at seven other schools that we know of; there may be more. It is easy to learn about these programs since they all have websites, listed in the Appendix. It is to be emphasized that these programs and their websites are set up and maintained primarily or exclusively by graduate students. To the best of our knowledge, all of these programs were organized primarily by people who had participated in Chicago's program, either as undergraduate mentees or as graduate student mentors. They are all based on the same model.

Graduate student initiative led to the DRP, which began in 2002. From the beginning it has been allied with the role of graduate students as stand-alone teachers in the lower level mathematics courses. The VCA began at the same time, also based on graduate student initiative, and the background is relevant. Chicago has an extensive and complicated system of employment for undergraduates as "Readers" (graders for honors calculus and higher level courses), "course assistants" (graders for the mid level calculus sequence) and "tutors" (teaching activities in the lower level calculus sequence).

When our VIGRE program started, our graduate student teachers were concerned that the course assistant job was less attractive than it should be and was not attracting a suitable caliber of grader. Their ideas for remedying that led to the institution of the VCA program. VCAs are not just graders. Rather they hold office hours and run independent problem sessions additional to those of the graduate student teaching the course. This program has led to markedly fewer complaints about graders from both the graduate student teachers and the undergraduates taking their courses. The increased collaboration of graduate students and undergraduates in the teaching of the lower level courses is one of the synergistic changes that has helped cement good relations between graduate students and undergraduates. The VCA position is now a permanent part of the undergraduate instructional program, funded internally, and we shall say little more about it.

2.2. How the DRP operates

Undergraduate mathematics majors want to see more mathematics, and our graduate students want to mentor them. The academic year DRP institutionalizes this. In this program, undergraduates meet one-on-one with graduate students to study some topic of mutual interest, agreed on between the student and mentor. The student learns all that he or she can. Participation is voluntary on the part of both undergraduates and graduate students.

Participants are required to meet with their mentor at least once a week during the quarter and to put in at least four hours of work on their topic per week. They are required to give a presentation at the end of the quarter. Projects must be approved by the committee. The program description advertises the following benefits of the program.

> Participating undergraduates will learn to work independently through studying a topic of their choice, well-suited to their interests. They will develop relationships with graduate student mentors and receive a good deal of personal attention focused on their mathematical studies. Finally, they will gain valuable experience in mathematical communication by giving a presentation on their work to an audience of their peers.

The last point refers to the end of quarter presentations. These are given at evening sessions, dinner provided by the department, attended by the undergraduate participants, their mentors, often some of the undergraduates and mentors who will be participating in the following quarter, and at least one faculty member. The level of the presentations has been uneven but generally very high.

The undergraduates who participate obtain no course credit and are not paid, and they do the work on top of their usually heavy course loads. Many of them are also working to help defray college expenses. The fact that more and more are participating, now averaging around twenty per quarter, argues that the advertised benefits are being delivered. The graduate student mentors are also volunteering their time. They receive some payment (currently $350 per quarter, plus up to $100 for books for the participants, with a little more for the organizing committee). Most other DRPs have a smaller number of participants and lower costs.

We emphasize that the choice of topic is arrived at by the participants themselves, mentee and mentor working together. The program is advertised by fliers around the department. Beyond that, recruitment is by word of mouth.

The program is run by a graduate student committee, under loose faculty supervision. Undergraduates apply to and are screened by the committee. Graduate students volunteer for the program, with recruitment of volunteers by the organizing committee. The committee sets up the mentorship pairings, taking into account the expressed interests in subject matter, perceptions of who would fit well with whom, and often input from those who have taught the undergraduates. The pairings are subject to faculty approval. The committee is self-sustaining, in the sense that its members recruit new members annually to ensure continuity. In fact, the faculty role is little more than attendance at the required student presentations and regular consultation.

3. The Summer REU

3.1. Background and expansion

This program has expanded steadily. Our VIGRE proposal requested 20 participants, and there were 22 in 2000, its first year of operation. At the time, even 20 participants seemed ambitious, especially since most REUs were (and are) considerably smaller. We did not intend a program as large as ours is today, and we had no precedent to guide us in scaling up. Quite a few relevant changes were suggested by the participants themselves, including some that led to central features of the current program. The program has evolved and is still evolving. The best piece of advice to anyone starting out as an organizer of any such program is to listen to the participants at all levels and learn from them.

Some years the program has had over 100 participants. Constrained by cost, the last few years have had around 80 participants, nearly all University of Chicago undergraduates, and that seems to be a reasonable steady state. It is natural to ask whether the restriction to University of Chicago students is an essential feature of the program. Opening the program to a few select outside participants is probably desirable and will be implemented in the next few years. However, anything like the present scale would be prohibitively expensive and logistically impracticable if opened to wide outside participation. For example, University of Chicago students find their own housing with negligible faculty or administrative input. Graduate students and others in the neighborhood seek to sublet apartments in the summer, and local websites facilitate this. Finding housing for comparable numbers of outside participants would be a nightmare. Other universities seeking to implement such a large scale program would surely face similar constraints.

3.2. Organization and finances

Students are free to follow their own interests. They choose courses, the topic of their research, and the exact mix of courses versus research based on their individual backgrounds and interests. They are assigned graduate student mentors, with whom they must meet at very least twice a week throughout the REU. There are many synergies between the regular academic year program and the REU. Some DRP academic year pairings continue into REU pairings, and some REU pairings continue into DRP pairings the following academic year. We will discuss the main aspects of the program, courses and mentored research, in the following sections.

Websites document the activities of the REU through the years. For the years 2002 through 2011, the address is

`http://www.math.uchicago.edu/~may/VIGRE/VIGREREU2011`

with 2011 changed to the year in question. For 2012–2014, the address is

`http://math.uchicago.edu/~may/REU2014`

with 2014 changed to the year in question. The more recent webpages have detailed course descriptions, hour by hour schedules, and information about the organization of the mentorship pairings. Undergraduate papers are posted on the annual websites, starting in 2006. Since 2007, they are separated into Full program and Apprentice program lists.[4]

There are other activities during the REU about which we will say little. There is an open house, mostly run by graduate students, to offer information and answer questions about going on to graduate study. However, there can easily be too much emphasis on future graduate study. To address this problem, the REU features a well-attended open house at which past participants working in non-academic jobs in the Chicago area answer questions and offer perspectives.

[4] The participant lists are so separated starting in 2004.

Maximum funding for a full eight week participant is $3,000. There is no additional funding for room and board. Apprentice participants receive just $1,000. Some students receive amounts in between, based in large part on their past academic performance and faculty recommendations. This is very far from ideal. It is a serious concern that we have no formal mechanism for determining need. Students are told that finances are not adequate to fully support all who deserve support. Those not needing support are urged to be generous, and some do choose to forgo support every year. Those in need of additional support are urged to tell the organizer, and every effort is made to make sure that few people, if any, do not participate only for financial reasons.

Graduate student mentors are expected to put in serious effort and are paid $2,000 for the full eight weeks, prorated for lesser periods of participation. While some faculty compensation is desirable, especially for postdocs teaching in the program, the twelve faculty participants in 2014 all taught pro bono. They understood how scarce funding is and were happy to volunteer. In past years, nearly all faculty participants received some small token of appreciation.

Even with that excruciatingly barebones level of individual support, the total cost of the program in 2014 is around $120,000. That may seem expensive, but the total cost per student, just $1,500, is surely lower than that of any other REU.

Aside from insufficient funding for the proper support of participants, perhaps the greatest defect of the REU has been that its organization and administration have been the work of a single faculty member. It is not nearly as difficult to set up such a program as one might imagine, but it does entail significant amounts of time. We emphasize that the actual operation of the program during the eight summer weeks is very much a cooperative effort and requires hardly any organizational or administrative work.

3.3. The 2014 REU

The program in 2014 had 81 participants, mentored by 27 graduate students (including three seniors going on to graduate study, who helped mentor in the apprentice program). Twelve faculty members taught in the program, five of them senior faculty, two assistant professors, and five Dickson Instructors. The high level of faculty participation is crucial. Participation is of course voluntary, and no great persuasion is needed. Here again word of mouth is critical. Postdocs often volunteer even before being asked to participate, having heard about the program from past participants. Senior faculty are often repeat participants. It is a lot of fun teaching students as committed as those participating in the REU!

There were over 130 applicants from the University of Chicago in 2014. It is hard to say no to an eager student at the home institution of a program like this, and some students have been allowed to sit in without funding. It is by now accepted that even permission to sit in is a privilege and that we just have to say no to many applicants. There were twelve sit-ins in 2014; they have exactly the same activities as the funded participants, including the requirement to write a paper. We have recruited a few African American and Hispanic students not from the University of Chicago, and that aspect of the program will undergo some future expansion. Recruitment so far has been ad hoc. The African Americans were recruited through the Mellon Mays and Leadership Alliance Programs; I recruited the Hispanic, who contacted me on his own initiative. We intend to learn how to recruit such people more systematically. Students recruited from outside are treated in exactly the same fashion as University of Chicago participants.

The program is only open to students who have just completed their first, second, or third years as undergraduates and who have taken at least the first year honors calculus sequence. (There are two tracks of calculus below that level and several more advanced tracks.) The program is advertised in all mathematics classes at the level of honors calculus or above. More precisely, the teachers in those classes, most of whom are postdocs, are told about the program and are asked to actively encourage their stronger students to apply. They also announce to their classes when applications become available. The undergraduate mathematics club also announces that. Word of mouth does the rest.

3.4. The apprentice program

One major innovation, forced by demand and the wish to welcome first year students, is a five week "apprentice program", intended primarily for first year students. It has been in operation since 2004, with just a few participants in 2004 and 2005. Formal implementation, with the apprentice program explicitly listed as an option on the application forms, began in 2006. This program has steadily increased in size ever since. There were forty apprentice participants in 2014.

A few of the more advanced first year students, five in 2014, participate from the outset in the full 8-week program. All apprentice participants are welcome to stay the full eight weeks, and in 2014 six did so. In recent years quite a few second year students, seven in 2014, have actually preferred to participate in the apprentice program. It serves as an entryway into mathematics for students with newly found interest in the subject.

There is a five week apprentice course, which is attended by all apprentice participants. It is nearly always taught by László Babai. It ranges over many topics, mostly in discrete mathematics of one sort or another. Since this course is so central to the apprentice program, here is the full 2014 abstract:

> The course will develop the usual topics of linear algebra and illustrate them on (often striking) applications to discrete structures. Emphasis will be on creative problem solving and discovery. The basic topics include determinants, linear transformations, the characteristic polynomial, Euclidean spaces, orthogonalization, the Spectral Theorem, Singular Value Decomposition. Application areas to be highlighted include spectral graph theory (expansion, quasirandom graphs, Shannon capacity), random walks, clustering high-dimensional data, extremal set theory, and more.

3.5. Courses in the 2014 REU

In addition to the apprentice course, more advanced courses are offered in a variety of areas. They range in duration from one to eight weeks and are taught by senior faculty and postdocs. They are frontloaded in the schedule. Although a few courses continue through the full eight weeks, most conclude earlier. Participants focus on individual work on their papers towards the end of the program; those papers are due by the end of August.

The content of the courses has varied widely from year to year. The more advanced subjects most regularly taught are topics in probability, geometry, topology, number theory, logic, and analysis, all of which had at least short courses in 2014. Abstracts of all courses are on the webpages, but we will give some idea of their content both to give an idea of the kind of mathematics the participants see and to indicate the breadth of faculty participation in the program.

In 2014, Gregory Lawler talked for the first two weeks on two related topics, "Random walk and the heat equation" and "Some challenging models in random walk". Lawler has written a book [1], aimed at undergraduates, that is based on his talks in the REU. Probability continued in weeks three through eight, taught by Antonio Auffinger, who is transitioning from an assistant professorship at Chicago to a tenure track position at Northwestern. His course focused on "Subadditive processes and an introduction to ergodic theory."

There were two consecutive and related three week geometry courses, taught by two Dickson Instructors, Dominic Dotterer and Sebastian Hensel. Two quotations from the abstracts give an idea. The first reads "What can listening to the harmonics of a graph tell us? We will explore ways in which the energy of the overtones quantify the geometric and topological complexity of the graph." The second reads "If a group acts nicely on a space we understand, then we can learn something about the algebra of the group."

Weeks one through four saw a course in number theory focusing on primes of the form $x^2 + ny^2$. It was taught by Matthew Emerton and two Dickson Instructors, Paul Herman and Davide Reduzzi. Like two other recent Chicago instructors, Herman was himself an undergraduate participant in the REU.

There were two one week courses in logic, one taught by Dennis Hirschfeldt on computability and definability and the other by Maryanthe Malliaris, an assistant professor, who managed to tell undergraduates about her celebrated recent solution, with Shelah, of the $p = t$ problem on cardinal invariants of the continuum.

There was a three week analysis course on "Equilibria in nonlinear systems" taught by another Dickson Instructor, Baoping Liu. It was coordinated with a new RTG summer program in analysis, described here:

`http://math.uchicago.edu/chicagoanalysis.`

Some advanced REU participants interested in analysis participated in both the REU and the analysis program. Its two week undergraduate component started a week before the REU, thus minimizing schedule conflicts.

Finally, there was an 8-week course on "Finite spaces and larger contexts", taught by May, that being the title of a book in preparation based on his presentations in the REU. Several research contributions by undergraduates have been based on topics in this area, which has the complementary virtues of being easily accessible, seriously interesting, and almost unknown even to experts in algebraic topology. It is a bridge between algebraic topology and combinatorics since finite T_0-spaces are the same thing as finite posets.

The idea behind this wealth of offerings is that students get some glimpse of what mathematics is really about, the great wealth of different directions it has to offer, and, perhaps above all, its interconnectivity. In the first few weeks of the REU, especially, students are encouraged to sample courses. Later in the program, they tend to attend only the courses of most interest to them, and especially courses related to their papers.

3.6. The required participant paper

A very major change in the REU was the institution of a required paper, beginning in 2006. This was a suggestion originally made by some of the undergraduate participants. It is now the heart of the program. Papers vary widely in scope, level, and purpose. Some contain original research but most, especially the apprentice papers, are primarily expository. They are written at the levels of the individual participants and vary widely in mathematical sophistication. Students know that their papers will be read and commented on by a senior faculty member, and they know that they will be posted online unless they request otherwise. They take the papers very seriously.

While students are encouraged to follow up questions and topics given during the courses and many of the papers are tied to one or another of the course offerings, others are not. The topics of the more advanced students often run far afield. Students are free to explore any direction, on their own and with mentors. We emphasize that students are urged to work at their own level of comfort.

The role of mentors is crucial. A few of the mentors may be new graduates on their way to graduate school (for the apprentices), a few may be postdocs and other faculty, but most are graduate students. The topics are chosen in consultation with the mentors, and the mentors are there to discuss all stages of the research and writing process. Papers must be submitted to the mentors and must be commented upon by them and revised accordingly before they are submitted to faculty. A senior faculty member reads and comments on all of the papers, and students are expected to revise them before they are posted on the web.

Just as with the DRP, the mentorship pairings are organized by a committee of advanced graduate students. However, in view of the larger scale and scope, these graduate students are recruited from year to year by the organizer of the REU, with whom they work in tandem. Both mentors and mentees report any problems to the committee (for example, lack of work by a participant or, more frequently, lack of expertise on the part of the mentor in the area of most interest to the mentee). Mentorship pairings are sometimes changed during the course of the summer as participant interests evolve. Mentors report at least weekly to the committee so that progress can be monitored and problems nipped in the bud, and the committee is in constant contact with faculty.

With 80 participants, individual faculty attention varies, but many of the more advanced participants are directly involved with faculty research. For example, one 2014 paper with four participant authors mentored by Babai finds "The smallest sensitivity yet achieved for a k-uniform hypergraph property", among other new results. To give an idea of the range of topics, we list the 2014 papers at the end. They run the gamut, including serious algebra, geometry, topology, analysis, probability theory, and logic. Some are sophisticated and give useful overviews and perspectives on particular fields. Others are playful. Some are elementary learning experiences. Some contain original research. We reiterate that students are urged to work at their own level of comfort, always mentored, but free to explore any direction. The amount of mathematics the students learn is prodigious. Most important, the students learn to love mathematics. They come to aspire to do mathematics. Many of the students spend two or three summers in the REU. Reading their papers from year-to-year and seeing their growth is eye-opening. Many current graduate students all over the country started out this way.

4. Graduate Student Participation; The Warm-Up Program

It will by now be obvious that the DRP and REU could not possibly function without dedicated graduate student cooperation. The VCA, DRP, and REU, described above as undergraduate programs are also graduate programs. The supervisors of the VCAs are the graduate students teaching the courses they assist in. The DRP is entirely organized and run by graduate students. The mentoring of undergraduates by graduate students is a central feature of the REU, and that is also organized and run by graduate students.

Graduate student participants in the REU are recruited by email requests for volunteers. This can be delicate. Usually the strongest of the graduate students are the ones who volunteer. However, some graduate students who would like to participate but are finding graduate study more difficult are better off focusing solely on their own research and are advised to do so. It is essential that graduate students genuinely volunteer, with no pressure on them to do so. Finding the right balance in the best interests of both the graduate students and the undergraduates they mentor has been fine tuned over the years. A ratio of three undergraduates per mentor, primarily grouped as six undergraduates under the mentorship of two graduate students, seems to work especially well. Other programs pay their graduate students more and work them harder.

Our programs could not function without an atmosphere of camaraderie and rapport among the graduate students themselves. That is developed from the beginning of their graduate student experience. In fact, it begins even before graduate students arrive. A graduate committee plays a huge role in the recruitment of graduate students, and virtually all graduate students participate in the entertainment of prospective students. In 2000, on their own initiative, graduate students instituted a warm-up program (WOMP) for the incoming class. It is now a permanent part of our program, funded by the Department of Mathematics. Graduate students give summary treatments of material that incoming students will be assumed to know during the first year program. Incoming students get acclimatized before classes begin. They get to know each other and they get to know the role graduate students play in undergraduate education.

The graduate student organizers of WOMP are recruited by faculty and past organizers working together. Essentially as part of their support package, entering students are offered $700 towards support during the two weeks of the warm-up program, and a total of $2500 is allotted to the organizers and the roughly twenty graduate students participating as hosts and speakers in the program.

A crucial feature of all of the programs we have mentioned is the special role that women graduate students play. A quarter of our graduate students are women. That is perhaps not a percentage to be proud of, but it is double the percentage of our peer departments. Virtually all of the graduate committees mentioned have women participants. They make an inestimable contribution to all aspects of the mathematics program.

5. The Balance of Learning and Research

The REU and the DRP both depend on close mentoring, and often joint learning, between undergraduates and graduate students. Students focus on learning and appreciating mathematics, often well beyond what is usually considered the undergraduate level. The students may or may not do research, but they get a taste of it, and they get a serious feeling for the world of graduate mathematics and beyond, serious enough to lead many of them to aspire to that as the way to spend their lives.

In both the REU and DRP, rigorous mathematics and a sense of community are combined. In both, the presentations are coupled with dinners and are social occasions. Students often bring their friends. Mathematics is many Chicago students' idea of fun, and it is amazing to see just how contagious a love of mathematics can be. One reason that it is so easy to recruit faculty and graduate students to participate in these programs is that they love the opportunity to teach serious mathematics, rigorously, to younger people who are genuinely interested in learning. The participants appreciate that they are seeing the real thing, mathematics that top flight research mathematicians consider to be important. The combination of a very high level of mathematics and respect for work at any level help to create the congenial atmosphere that makes these programs so popular.

What undergraduate research has been undertaken in these programs? There has been some, and some of it has been seriously interesting, but in truth there has not been a great deal in comparison with the number of participants. There is a statement from a 1961 conference on undergraduate research at Carleton College to the effect that "The aims of undergraduate research are the training and stimulation of the student, not the attainment of new results, though such bonuses will come occasionally." I must admit to being a retrograde iconoclast: that does accurately portray the attitude at Chicago. We are very happy to give some of our undergraduates research topics. We are delighted when they prove something new and interesting. We do rope them in by having them work on things we would like to better understand ourselves.

But Chicago is a premier research department. To achieve the level of postdoc and faculty commitment that we have from people at the top of their game, we naturally give complete leeway in what and how they present mathematics to the participants. If they want to focus on undergraduate research, they are free to do so. But they generally don't. To reiterate, ours is not a top-down program. We don't tell postdocs, let alone tenured faculty members, how they should run their REU classes. They can and do run them as they please.

It is the most enthusiastic teachers who volunteer, and their energy and enthusiasm are central to the success of the program. We are not focused on undergraduate research per se. We are focused on attracting many of the young people in our care to careers in mathematics and the sciences, and attracting all of them to make mathematics a part of the rest of their lives.

6. Some Quotes from 2014 Participants

Here are some quotes from 2014 participants. (More quotes may be found in [2].) They are from gratuitous thank yous in emails to which their 2014 papers were attached or from acknowledgments in those papers, and this is just a sampling. I apologize if their inclusion seems immodest. The point is to illustrate the joy people take in the program, which explains why the popularity of the program is contagious. That is the crucial reason the program seems likely to continue.

"Thank you very much for a fantastic Math REU. I gained a new appreciation for high-level mathematics this summer."

"Thank you for running this program; I had a great time, and being completely immersed in math was a fantastic, albeit at times overwhelming, experience."

"Here is my final REU paper. Thank you very much for allowing me to participate and gain meaningful exposure to different disciplines through this program."

"My REU paper is attached. Thank you for a fantastic summer!"

"Thanks so much for all the work you did organizing the program, and for the work you will do reading everyone's papers! I really appreciated the opportunity to learn so much math with so many brilliant people!"

"Thank you for the work you put in to organize the program this year. Over the course of the program, I learned a lot of interesting math and was challenged to think deeply about math and to express my thinking in ways that I have not been before. I feel very fortunate to have been able to participate."

"Thank you so much for running this REU program; I honestly feel like I've learned more math this summer than I ever have, and I would not have wanted to spend my summer doing anything else. It has been a fantastic experience, and I really hope you enjoy my paper!"

"I would also like to thank — for coordinating this wonderful undergraduate research and study opportunity."

"I would like to thank — for making possible the most mathematically productive eight weeks of my life."

"Thanks so much for running the REU this year! It was truly a mathematical blessing."

From an apprentice participant: "Thank you so much for giving me the opportunity to participate in such wonderful program this summer and it would be great if I could return next summer for the full program."

"I also would like to thank — and the University of Chicago Math Department for making this experience possible. It really was wonderful."

"I want to thank — and all of the instructors in the math REU for this wonderful program that helped me to both broaden my knowledge of mathematics and increase my mathematical maturity."

7. Titles of 2014 Participant Papers

FULL PROGRAM PAPERS

- Calista Bernard. Regularity of solutions to the fractional Laplace equation.
- Joshua Biderman, Kevin Cuddy, Ang Li, and Min Jae Song. On sensitivity of k-uniform hypergraph properties.
- Ben Call. Introduction to Furstenberg's 2x3 conjecture.
- Zefeng Chen. Quasi-preference: choice on partially ordered sets.
- Sean Colin-Ellerin. Distribution theory and applications to PDE.
- Kevin Cuddy. See Joshua Biderman.
- Kyle Gannon. Introduction to the Keisler order.
- Claudio Gonzales. Polynomials in the Dirichlet problem.

- Justin Guo. Analysis of chaotic systems.
- Jackson Hance. Hodge theory and elliptic regularity.
- Jordan Hisel. Addition law on elliptic curves.
- Yifeng Huang. Characteristic classes, Chern classes and applications to intersection theory.
- Sofi Gjing Jovanovska. Beck's theorem characterizing algebras.
- Sameer Kailasa. Topics in geometric group theory.
- Simon Lazarus. Basic algebraic geometry and the 27 lines on a cubic surface.
- Fizay-Noah Lee. Kummer's theory on ideal numbers and Fermat's last theorem.
- Ang Li. See also Joshua Biderman. The Lefschetz fixed point theorem and solutions to polynomials over finite fields.
- Siwei Li. Strategies in the stochastic iterated prisoner's dilemma.
- Jason Liang. Measure-preserving dynamical systems and approximation techniques.
- Lucas Lingle. Intro to class field theory and the Chebotarev theorem.
- Ben Lowe. The local theory of elliptic operators and the Hodge theorem.
- Benjamin McKenna. The type problem: effective resistance and random walks on graphs.
- Redmond McNamara. Introduction to de Rham cohomology.
- Jing Miao. Convergence of Fourier series in L^p space.
- Victor Moros. The zeta function and the Riemann hypothesis.
- Sun Woo Park. An introduction to dynamical billiards.
- Nicholas Rouse. Compact Lie groups.
- Bryan Rust. A theorem of Hopf in homological algebra.
- Maximilian Schindler. Basic Schubert calculus.
- Noah Schoem. Well-foundedness of countable ordinals and the Hydra game.
- Joel Siegel. Expander graphs.
- Maria Smith. Kolmogorov–Barzdin and spacial realizations of expander graphs.
- Min Jae Song. See Joshua Biderman.
- Blaine Talbut. The uncertainty principle in Fourier analysis.

- Victor Zhang. Incompleteness in ZFC.
- Yuzhou Zou. Modes of convergence for Fourier series.

APPRENTICE PROGRAM PAPERS

- Will Adkisson. Geodesics of hyperbolic space.
- Fernando Al Assal. Invitation to Lie algebras and representations.
- John Alhadi. Exploration of various items in linear algebra.
- Kevin Barnum. The axiom of choice and its implications.
- Karen Butt. An introduction to topological entropy.
- Rachel Carandang. Generalization in machine learning: the Vapnik–Chervonenkis inequality.
- David Casey. Galois theory.
- Spencer Chan. Compass and straightedge applications of field theory.
- Cindy Chung. An introduction to computability theory.
- Brenden Collins. An introduction to Lie theory through matrix groups.
- Paul Duncan. The Gamma function and the Zeta function.
- Adam Freymiller. Markov chain tree theorem and Wilson's algorithm.
- Michael Hochman. Chutes and ladders.
- Yunpeng Ji. Discussion of the heat equation.
- Ken Jung. Brouwer's fixed point theorem and price equilibrium.
- Josh Kaplan. Binary quadratic forms, genus theory, and primes of the form $p = x^2 + ny^2$.
- James W. Kiselik. Conic and plane curves.
- Daniel Kline. The structure of unit groups.
- Stefan Lance. A survey of primality tests.
- Scarlett Li. Brouwer's fixed point theorem: the Walrasian auctioneer.
- Jack Liang. Rudimentary Galois theory.
- Larsen Linov. An introduction to knot theory and the knot group.
- David Mendelssohn. Operations and methods in fuzzy logic.
- Matthew Morgado. Modular arithmetic.
- Seth Musser. Weakly nonlinear oscillations with analytic forcing.

- Adele Padgett. Fundamental groups: motivation, computation methods, and applications.
- Daniel Parker. Elliptic curves and Lenstra's factorization algorithm.
- Robert Peng. The Hahn–Banach separation theorem and other separation results.
- Peter Robicheaux. Calculation of fundamental groups of spaces.
- A very Robinson. The Banach–Tarski paradox.
- Zachary Smith. Fixed point methods in nonlinear analysis.
- Aaron Geelon So. Symbolic dynamics.
- Matthew Steed. Proofs of the fundamental theorem of algebra.
- Ridwan Syed. Approximation resistance and linear threshold functions.
- Shaun Tan. Representation theory for finite groups.
- Zachry Wang. Itō calculus and the Black–Scholes option pricing theory
- Nolan Winkler. The discrete log problem and elliptic curve cryptography.
- Eric Yao. Plane conics in algebraic geometry.
- Joo Heon Yoo. The Jordan–Chevalley decomposition.

Appendix: Websites of DRP Programs

The University of Chicago
`http://math.uchicago.edu/~drp/rails`
The University of California, Berkeley
`http://math.berkeley.edu/wp/drp`
University of Connecticut
`http://www.math.uconn.edu/degree-programs/undergraduate/directed-reading-program/`
The University of Maryland
`http://drp.math.umd.edu/`
MIT
`http://math.mit.edu/research/undergraduate/drp/index.php`

Rutgers University
`http://www.math.rutgers.edu/undergrad/Activities/drp`
The University of Texas
`http://www.ma.utexas.edu/rtgs/geomtop/DRP.html`
Yale University
`http://math.yale.edu/directed-reading-program`

References

[1] G. F. Lawler, *Random Walk and the Heat Equation.* American Mathematical Society, 2010.

[2] J.P. May, The University of Chicago's VIGRE REU and DRP, in ed. J.A. Gallian *Proceedings of the Conference on Promoting Undergraduate Research in Mathematics.* American Mathematical Society, 2007, pp. 335–340.

[3] S. Schuster and K.O. May, *Summary and Resolutions, Carleton 1961 Conference.* Reprinted in L.J. Senechal ed., *Models for Undergraduate Research in Mathematics.* MAA Notes No. 18, Mathematics Association of America, 1991, pp. 157–159.

Chapter 7

Why REUs Matter

Carlos Castillo-Garsow
Eastern Washington University, Cheney, WA, USA
cccastillogarsow@ewu.edu

Carlos Castillo-Chavez
Arizona State University, Tempe, AZ, USA
ccchavez@asu.edu

1. History and Introduction

The Cornell-SACNAS Mathematical Sciences Summer Institute (CSMSSI) was a mathematical biology research experience for undergraduates (REU) founded in 1996. One year later, the program was renamed as the Mathematical and Theoretical Biology Institute (MTBI). In 2004, MTBI moved to Arizona State University (ASU), where it merged with a K-12 program, the Institute for Strengthening the Understanding of Mathematics and Science (SUMS). In 2008, MTBI/SUMS expanded again to become embedded in what is now the Simon A. Levin Mathematical, Computational and Modeling Sciences Center (MCMSC). The purpose of the larger center is now to connect MTBI's education-through-research mission directly to undergraduate and graduate programs in the mathematical sciences and to cutting edge research activities at the interface of the mathematical, life, and social sciences.

Now, in our 20th year, MTBI/SUMS has been extraordinarily successful by most traditional measures. From 1996 through its 2014 summer program, MTBI has recruited and enrolled a total

MSC 2010: 01A65, 01A72 (Primary); 01A80 (Secondary)

of 423 regular first-time undergraduate students and 91 advanced (returning) students. MTBI students have been prolific researchers, with 180 technical reports,[1] and a large number of refereed publications (including but not limited to the following representative publications, [14–17, 19, 24–28, 31, 33, 36, 37, 40, 43, 46]).

MTBI has also been successful in recruiting and retaining students in mathematical sciences. Through December 2014, 249 out of 357 (70%) of US MTBI student participants had enrolled in graduate or professional school programs and 109 MTBI student participants had completed their PhDs. 74 students have received their PhDs since 2008, for a PhD. completion rate of a little over 10 PhDs per year.

MTBI has received funding from Cornell University, Arizona State University, Los Alamos National Laboratory, the Sloan Foundation, the NSA, and NSF. NSF and NSA currently fund MTBI each at roughly 120K per year, with graduate mentors supported by teaching and research assistantships (127 since 1996) mostly funded via university funds and Sloan fellowships.

The program has also received external recognition in the form of multiple national awards. The Director of MTBI was awarded the Presidential Award for Excellence in Science, Mathematics and Engineering Mentoring (PAESMEM) in 1997 and the American Association for the Advancement of Science Mentor Award in 2007. Also in 2007, the AMS recognized MTBI as a Mathematics Program that Makes a Difference. MTBI's partner high-school program, SUMS, was recognized and awarded the Presidential Award for Excellence in Science, Mathematics, and Engineering Mentoring in 2002, and MTBI received the same award in 2011.

The reason for MTBI's extraordinary success is that the primary goal of MTBI has always been to effect social change. MTBI/SUMS's mission is to encourage students — especially women, Chicano/Latino, Native American, and African American students — to pursue advanced degrees in the computational and mathematical sciences, with particular emphasis in the applications of mathematics to the life and social sciences. 249 (70%) of MTBI

[1] http://mtbi.asu.edu/research/archive.

undergraduates are underrepresented minorities and/or members of the Sloan Pipeline Program (Underrepresented minorities (URMs) include Hispanic, African-American and Native American students). Since January 1995, 69 URMs (77%) holding US citizenship or permanent residency have completed PhDs, and 51% of all MTBI PhD recipients are women. Research, funding, and external recognition are means to pursue the ultimate mission of social change, and the mission of social change results in quality research, funding, and recognition. Students — particularly under-represented students from impoverished backgrounds — have a keen interest in social change. Faculty are inspired by this mission, and many guest faculty participate frequently, voluntarily and without pay.

In short, we have built and continue to build a community model to start new undergraduate majors and PhD degrees over the past 19 years — its effectiveness is tied to the spirit of community, volunteerism, and service that pervades every aspect of the program. From this perspective of effecting social change through mathematical research, MTBI has succeeded, but there remains much to do. It is this ongoing mission and the continued sense of dissatisfaction with MTBI's successes to date that has driven MTBI faculty and staff to continue to do better, to continue to grow over the past 19 years from a simple summer REU to a nationally recognized pipeline program that tracks and supports under-represented students from high-school to tenure.

In particular, ever since the merger with SUMS in 2004, MTBI/SUMS has become increasingly interested in K-12 education. The addition of SUMS and partnership with other high-school programs in Arizona have improved our capability to serve the URM population by giving us the ability to grow URM talent for MTBI, rather than rely solely on the recruitment of students who make it to college on their own — similar to the way that the REU grows talent to be recruited to graduate school.

However, this partnership has also made us aware of other ways in which K-12 partnerships can be mutualistic. A short example of this type of collaboration is the Center's newly funded initiative to reintroduce a teaching and mentorship component to our programs. MTBI will be collaborating with Dr. Raquell Holmes of ImproviScience to provide educational research-based workshops on mentoring and collaboration for both student mentors and faculty.

More generally, we are proposing two more initiatives to strengthen mutually beneficial ties between mathematical science REUs and communities of educators. We would like other REUs to be inspired by the goal of social change and join us in these or similar initiatives. We intend to show how REUs are uniquely positioned to improve the effect of social change by collaborating with mathematics education researchers and by recruiting from the population of qualified teaching majors. We will do this using examples from mathematics education research as well as from MTBI/SUMS itself; and in the process we hope to give the reader a more detailed picture of how MTBI works, the mechanics of its successes to date and, we hope, the desire to have a dialogue that help us to improve what we do by learning first-hand what other groups have done so well over several decades.

2. The Mechanics of MTBI/SUMS

There are a number of factors that play a role in the educational success of MTBI/SUMS. The most important seem to be the mathematical content, the reversal of hierarchy, and the community model.

MTBI/SUMS summer sessions are divided into two parts. For the first three and a half weeks, students attend lectures and do homework (*the mathematical content*). For MTBI, this content includes difference and differential equations, statistics, stochastic processes, agent based modeling,[2] and computer simulation. The program roughly follows the text of Brauer and Castillo-Chavez [4], supplemented by guest lectures. For SUMS, this content is typically difference equation modeling only, based on an idea from Robert May [30].

The second half of the program is student-driven group research projects. In MTBIs initial prototype year (1996), the projects were

[2]Models of large numbers of individuals acting according to pre-determined algorithms; usually, but not always simulated via computer.

assigned as tasks. However, this did not generate the desired student buy-in or the experience of doing authentic research. In subsequent years, students were expected to design their own projects, while returning students, graduate students and faculty served in advising roles. Because students choose their topic of study, they frequently know more about the situation than the mentors (*flipped hierarchy*). The mentors contribute mathematical and modeling experience, but rarely situational knowledge [6, 7, 11].

Over the entire program, the faculty deliberately work to create a research community (*the community model*). MTBI is a residential program, and during the first half of the program, participants are deliberately given far more work than they can complete on their own, forcing them to make use of their peers. Even in groups, students typically work 10–12 hours per day, six days a week throughout the eight weeks. In order to explicitly encourage collaboration, students are organized and participate in weekly group meetings. These groups then form the foundation of the collaborative research project. On alternating weekends, students participate in organized activities such as day trips, water parks, a Fourth of July celebration and other social outings. These activities build esprit-de-corps among the students and facilitate their teamwork in research. Students are also encouraged to return as advanced students, graduate students, post-docs, and faculty — building long-term bonds with the research community and serving as role models for new students.

The dynamics of the community model of MTBI have been discussed in detail elsewhere [11, 17]. So in this article, we will discuss the importance of the other two factors: the mathematical content and the reversal of hierarchy.

3. The Reversal of Hierarchy

Research modeling experiences form the keystone of the MTBI/SUMS summer experience. The entire program is designed to prepare students to successfully pursue a research modeling experience. Students are trained in mathematical biology techniques, but in the final half of the session, they apply these techniques to their own curiosities. In MTBI, this has led to final projects in a variety of diverse topics including but not limited to: the three strikes law [42], gang recruitment [1], education [3, 20], immigration [13], political third party formation [38], mental illness [18, 21, 25], pollution [5], obesity [22], drug use [32, 43], and even MTBI itself [17]. Students choose these topics because they are personally meaningful. Nearly every one of these off-topic applications is chosen because a group member has personal experience with the problem. Either they themselves, or a family member, or a close friend has run afoul of a gang, or has dropped out, or has struggled with a mental illness. This personal experience both motivates the group and supplies them with valuable insider expertise.

Because MTBI students choose their own research projects, they frequently know more about the topic than their graduate student and faculty mentors. This creates the "reversal of hierarchy" in which the students take the leadership role in the project, and the mentors serve as consultants. As consultants, the mentors provide missing mathematical expertise to judge the feasibility of the project, suggest appropriate tools, and tutor the students in any additional techniques they need for their project that were not covered in the general lectures.

SUMS students have similar experiences, but the mentor role is even more involved. Consider the most recent SUMS year, in which the students were trained in difference equation techniques for mathematical biology. Their interests, however, were diverse, and they did projects in economics, education, driving safety, and zombie epidemiology. In order to complete these projects, some student teams required additional mentoring in statistical and agent-based programming tools not originally covered by the difference equation curriculum.

This student leader with consulting mentor approach is a critical factor to the success of MTBI/SUMS. Without students choosing their own projects, a vital portion of the research experience is missing, and students are not truly doing research. Without the mentor in a consulting role, projects quickly exceed both the mathematical abilities of the students and the time they have available to complete the project. One of the key roles of the mentor is helping students slim their interests down to a project that can be completed in four weeks.

4. The Importance of Students Doing Research

Making the student the leader of the research is vital because it makes the mathematical experience real. When students are in the leadership role, they must draw on their own experiences to make sense of mathematics and to make decisions. Contrast this with experiences that students typically have in modeling in K-12 schools.

Reusser [35] describes the story of 97 first and second graders who were asked the following nonsensical questions:

> "There are 26 sheep and 10 goats on a ship. How old is the captain?"
> "There are 125 sheep and 5 dogs in a flock. How old is the shepherd?"

76 of those 96 first and second graders were able to provide a numerical solution to these questions, by finding an appropriate operation for the numbers that would result in an age. In the first task adding to 36 years old, and in the second task dividing to 25 years old.

The problem stems from the social norms that are quickly established in a mathematics classroom. Every question must have an answer, and that answer must be reached using the tools that are currently being studied. Or, as Reusser put it: "Almost every systematic dealing with ambiguity and unsolvability is factually excluded from textbooks, from curricula, and from the school setting where it even seems alien."[3]

4.1. Pseudocontext

It should not surprise any of us that students learn that mathematics is not about making sense. Textbooks have long histories of including "modeling" exercises that do not require students to make sense, in some cases the tasks are actively hostile to making sense. Jo Boaler

[3]For more cases and anecdotes on lack of sense making in school mathematics, see [41].

coined the term "pseudocontext" for referring to these types of tasks, saying:

> A restaurant charges \$2.50 for $\frac{1}{8}$ of a quiche. How much does the whole quiche cost?
>
> ...
>
> Everybody knows that people work together at a different pace than when they work alone, that food sold in bulk such as a whole quiche is usually sold at a different rate than individual slices, and that if extra people turn up at a party more pizza is ordered or people go without slices — but none of this matters in Mathland. [2, p. 52]

If students were required to make sense of the quiche task in the context of their real world experience of purchasing quiche, then the quiche task [2, p. 52] is just as unsolvable as a captain problem from Reusser [35]. The quiche task is only "solvable" because of the social norms established in the classroom. The question is being asked in a math class, and a question in a math class must have a solution, therefore the normally incorrect assumption that the price of quiche is the sum of the price of its slices must be valid here. The same solvability assumption that allows students to answer the quiche task "correctly" is also what leads the majority of students to solve the captain task incorrectly.

4.2. Research modeling

Contrast this type of assigned task with the experiences of SUMS students, or with the following case from Resnick [34, pp. 68–74]. Resnick describes two high school students, Ari and Fadhil, who were working with the agent-based modeling program StarLogo. At the same time, they were enrolled in driver's education. Ari and Fadhil developed a curiosity: they wanted to know what caused traffic jams. Using StarLogo, Ari and Fadhil developed several simulations of drivers on a highway, and explored driver behaviors that contributed to or eliminated traffic jams. Although Ari and Fadhil did not succeed in controlling their simulated traffic jam, they discovered quite a bit about traffic jam behavior: that traffic jams moved as waves in the direction opposed to traffic, and that starting all cars at the same initial speed did not prevent a traffic jam, so long as the cars were unevenly spaced.

Ari and Fadil's story differs from the quiche example in two critical ways: first, there was no externally imposed task. Instead, Ari and Fadil were pursuing their own curiosities. Secondly, Ari and Fadil were not looking for a solution to be validated by an authority figure. They were looking for the understanding that would satisfy their curiosity. Taken together, we have two high schools students pursuing a non-mathematical curiosity with mathematical tools while believing that no pre-existing solution existed. In other words, Ari and Fadil were engaged in mathematical research. This is exactly the sort of activity that quiche tasks do not prepare students for.

Based on the experiences of MTBI/SUMS students, as well as Ari and Fadhil's example, we propose three criteria for identifying research modeling activity:

1. The problem is based on the student's non-mathematical experience.
2. (Therefore) The problem originates with the student.
3. The goal of the activity is understanding, not a solution.

5. Scaling up to K-12

It should seem obvious that implementing research modeling experiences in K-12 education is desirable, both as a tool for recruiting students to STEM fields and training tool for preparing students for work in these fields. However, with the current state of teacher training, it is not feasible beyond a limited scale. There are small-scale programs that work on exposing K-12 students to modeling experiences, including SUMS, the St. Laurence County Mathematics Partnership at Clarkson, the Center for Connected Learning at Northwestern, or StarLogo at MIT. However, these programs revolve around the expertise of a PhD applied mathematician or graduate student doing the leading and the mentoring. There seem to be three barriers to implementation modeling research activities on a broad scale in K-12: First, students are not mathematically prepared to engage in these types of activities; second, teachers are not mathematically agile enough to mentor them; and third, teachers do not have the experience to even imagine these sorts of activities in the first place.

5.1. Preparing MTBI/SUMS students for research

All REUs prepare students to do research, and an examination of these programs can reveal ways to prepare students to do research that can be scaled up to a larger population. Again, we use MTBI/SUMS as an example.

In order for a reversal of hierarchy to be successful, students must first be in a position where satisfying a curiosity mathematically seems natural. Students must be mathematically prepared so that they can be successfully inspired to think of the types of curiosities that are well satisfied by a mathematical approach, and so that they can pursue those curiosities with enough mathematical competence that mentoring is feasible.

The lecture portion of MTBI/SUMS makes use of a third type of modeling activity. In the middle of the spectrum between pseudocontext and research modeling lies the modeling exercise, exemplified by these problems from the MTBI and SUMS homework assignments:

1. (SUMS 2014) Using your favorite method, find all equilibria of the model (you can assume λ is a positive constant):
$$P_{t+1} = \frac{\lambda P_t}{(1+P_t)^2}.$$
2. (SUMS 2014) Say we are studying a nonlinear model made up of predators (P) and prey (Q), where $-sPQ$ denotes the deleterious effect P has on Q, and kPQ represents the positive effect of Q on P. What is the biological interpretation of the assumption that $s > k$?
3. (MTBI 2014) Limpets and seaweed live in a tide pool. The dynamics of this system are given by the differential equations
$$\frac{ds}{dt} = s - s^2 - sl,$$

$$\frac{dl}{dt} = sl - \frac{l}{2} - l^2,$$
$$l \geq 0, \quad s \geq 0.$$

(a) Determine all equilibria of this system.
(b) For each nonzero equilibrium determined in part (a) evaluate the stability and classify it as a node, focus, or saddle point.
(c) Sketch the flows in the phase plane.

These modeling exercises have some similarity to pseudocontext tasks. Similar problems — in which a context is provided along with an equation and then the student is asked to work only with the equation — are a staple of "application" problems in nearly every textbook. However, MTBI/SUMS modeling exercises are demonstrably successful in preparing students for research modeling while other superficially similar exercises are not.

Unlike the research modeling activities described above, these are much more structured exercises that target the development of specific mathematical tools used in studying dynamical systems. Although the models provided in these examples have biological meaning, the student is not always asked to interact with the problem in context. In the above examples, the student is only required to use the context in Example 2. Examples 1 and 3 only require that the student interact with the task mathematically. The context could be completely excluded and the task could still be solved. We suspect that the differences between these exercises and pseudocontext are not so much in the individual exercises, but rather in their use: their place in the larger MTBI/SUMS program.

Every student in MTBI/SUMS knows that the session will end with an assignment to do their own project. In this scenario, these modeling exercises are embedded to take on a different significance. While the context may not be mathematically necessary to solve the exercise in front of them, it is necessary to prepare them for their projects. The context becomes an example of the types of modeling situations that can be addressed with this particular mathematical approach. Students know that they will need these examples to pursue their own projects. So it is not the individual excercise that is important for the goal of teaching a modeling perspective, but rather the accumulation of exercise that shows what types of problems can be addressed mathematically, and what different types of scenarios lend themselves to different approaches.

In addition to situational context, MTBI/SUMS modeling exercises must also be examined in their mathematical context. An isolated exercise may just be about applying a particular mathematical technique, such as finding a Jacobian. In the larger context, however, exercises are selected so that students will experience critical distinctions in dynamical systems: between discrete and continuous, linear and nonlinear, deterministic and stochastic. Each distinction affects both the behavior of the system, the degree to which the system can be explored, and the tools used to explore it. It is in this area that teacher training is lacking. They rarely have the experiences necessary to emphasize these distinctions in their own mind.

5.2. Attending to teacher' training

Teachers tend to teach mathematics in the way they were taught, and the mathematics that teachers learn in school and teach in school is deficient for preparing students for mathematical research. Exploring the ways in which REUs prepare students for mathematical research can highlight areas of school mathematics that need changing, and areas of teacher training that can be improved. Consider the following examples based on MTBI's highlighted distinctions: between discrete and continuous, linear and nonlinear, deterministic and stochastic.

5.2.1. *Discrete and continuous*

The distinction between discrete and continuous has a huge impact on the behavior of dynamical systems. While a one-dimensional discrete system such as $P_{n+1} = rP_n(1-P_n)$ can exhibit chaotic behavior, chaos is impossible in continuous systems of fewer than three dimensions, no matter how complex. However, work with students has shown that the distinction between discrete and continuous can be very difficult for students who are used to exercises in plotting points and connecting the dots [29]. Even highly successful students will show a preference for whole number counting and regularly spaced rational numbers that interfere with their ability to draw conclusions about continuous systems [9, 10, 12], and secondary teachers suffer from similar interference from discrete thinking [45].

5.2.2. *Linear and nonlinear*

In traditional K-12 schooling, "linear" is a modifier that describes primarily functions or graphs of functions, so linear means "straight" and nonlinear means "curved." The meaning of linear becomes slightly extended when classes begin to discuss "systems of linear equations," but a "linear equation" really only means that each variable can be expressed as a linear function of the others, that is, "linear," in this sense, is still modifying function. In general, K-12 students and teachers deal with models that depend on linear concepts and so learning the difference between linear and nonlinear problems is challenging particularly at the K-12 level.

Robert May [30], uses "linear" as a modifier not only for "linear functions" or "linear equations," but also for "linear systems" and "linear problems," a critical distinction when the goal is to study dynamics. May distinguishes between linear or nonlinear systems of differential equations and linear or nonlinear difference equations — the principle of superposition being in general lost in the nonlinear world. For example, in the equation $X_{t+1} = aX_t$, X is a nonlinear (more precisely geometric) function of t, but the equation is a "linear equation" because X_{t+1} is a linear function of X_t; the model is based on linear concepts. In contrast $X_{t+1} = aX_t(1 - X_t)$ (*), "arguably the simplest interesting nonlinear difference equation," with X_{t+1} a nonlinear function of X_t, is not based on linear concepts and as a consequence, here we lose superposition. May sees this distinction as critical saying:

> The elegant body of mathematical theory pertaining to linear systems (Fourier analysis, orthogonal functions, and so on), and its successful application to many fundamentally linear problems in the physical sciences, tends to dominate even moderately advanced University courses in mathematics and theoretical physics. The mathematical intuition so developed ill equips the student to confront the bizarre behaviour

> exhibited by the simplest of discrete nonlinear systems, such as equation [(*)]. Yet such nonlinear systems are surely the rule, not the exception, outside the physical sciences. [30]

In his classic *Real and Complex Analysis*, Walter Rudin introduces the exponential function as the most important function in mathematics [39, p1]. The exponential family is typically defined as the unique solution of the linear problem $\frac{dx}{dt} = ax, x(0) = x_0$, that is, when the rate of change of a function is proportional to the value of the function itself.[4] The solution of this linear differential equation is by definition the exponential function. Part of the importance of the exponential function lies in its ability to "locally" approximate nonlinear systems; no different than using tangent planes to approximate surfaces locally, a process referred to as "linearization" or "linear analysis" [30], intimately connected to the principle of superposition.

Robert May [30] has suggested that this distinction between linear or nonlinear systems or problems (concepts based on the study of dynamics) be introduced to students as early as possible in their education, before calculus. Examples such as SUMS show that this suggestion can be implemented realistically, and further experiments show that the linear property (rate proportional to amount) used to define an exponential function can be realistically introduced before calculus as well [10, 44].

5.2.3. *Deterministic and stochastic*

Lastly, stochastic processes (such as agent-based models) have been shown to have tremendous applications in helping students develop a research modeling mindset [34]. Because the high degree of complexity can be managed by computer experiment, a little programming training allows students to explore complex systems they would not be mathematically equipped to handle with algebraic tools. They can be thought of as a "gateway drug" into mathematical modeling: a gentle beginning that encourages students to try harder math later on. Agent-based models and Markov chains (both discrete

[4]Generalizations include, for example, $\frac{dX}{dt} = AX$ where A is an $n \times n$ matrix and X is an $n \times 1$ vector.

and continuous) are popular techniques among the students of both MTBI and SUMS.

5.2.4. *Suggestions*

Here, REUs can have a social impact by inviting mathematics education researchers to observe and study. Mathematics education researchers can take what they learn about effective preparation for research from REUs and use these results to improve teacher training programs, or to inspire further research into high school student learning. In fact, much of the research cited above [8–10, 12] was born from collaborations between a mathematics education researcher and MTBI. This example shows the possibility of such fruitful collaboration.[5]

5.3. Attending to teachers' experiences

However, the greatest barrier to the implementation of mathematical modeling research activities in schools is simply that very few teachers can imagine them. Teachers rarely attend REUs and rarely attend graduate school in applied mathematics. The simple fact is the vast majority of people charged with teaching mathematical modeling have never done it.

Without some sort of experience in mathematical modeling, teachers cannot imagine what it is, let alone how to implement it in their own classrooms. Instead, they are left to rely on the experiences that they do have: textbook exercises and pseudocontext examples in which the applications are informed, but not genuine or meaningful. This cycle then perpetuates itself as future teachers grow up and learn in exactly the same limited environment.

Here again, REUs can have an impact. Teachers are trained as undergraduates. Secondary teachers in particular receive extensive mathematical training that qualifies them for REUs, and REU programs can provide exactly the mathematical research experience that teachers need in order to begin imagining a different sort of task for

[5] For more on the role of example cases in science, see [23].

their students. REUs do not generally recruit from this pool because teachers are rarely interested in graduate study, and not all REUs will be interested or even appropriate for this sort of work; but for REUs that are interested in effecting social change, recruiting qualified undergraduate mathematics education majors is one possible way to having a large impact.

6. Recommendations and Future Work

REUs have tremendous potential for social impact. MTBI is one example of such a program, one which has been tremendously successful because of its social agenda. Not all REU directors will be interested in taking this route, and that is understandable. But for REU directors who wish to use their program to effect social change, we have some recommendations. Improving K-12 education is an area of high need, and this is an area where social impacts can be made with only small changes to an REU program, or more sweeping changes if desired.

First: build collaborations with mathematics education faculty. Collaborations between MTBI/SUMS and mathematics education faculty have been fruitful, resulting in improved understanding of how the REU operates, well received mentorship workshops for REU faculty and graduate students, and improvements to teacher training programs run by the mathematics education faculty. Building a collaboration with a mathematics education researcher need not be elaborate or time consuming. It is simply a matter of issuing an invitation to the right person.

Second: recruit secondary mathematics education majors. Future teachers can have a huge impact. A successful teacher will teach upwards of 5000 students in their lifetime. Creating good mathematical modeling experiences for 5000 students improves mathematical citizenship for everyone, and increases interest in STEM major programs. Recruiting and contributing to the training of several future teachers have the potential to dramatically change the landscape that undergraduate programs recruit from. Here again, we do not suggest that REUs make dramatic changes to their program. Secondary mathematics education majors are qualified for these programs by their coursework, and adjusting the curriculum would eliminate the authenticity of the research experience. Instead, we suggest that REU faculty simply make an effort to invite these students (particularly strong students that the faculty personally encounter in their teaching), and market to the demographic. REU development does not happen all at once. Similar to the way MTBI has changed over decades, programs can scale up as the needs of these students are better understood through experience.

Acknowledgments

This project has been partially supported by grants from the National Science Foundation (DMS- 1263374 and DUE-1101782), the National Security Agency (H98230-14-1-0157), the Office of the President of Arizona State University (ASU), and the Office of the Provost of ASU.

References

[1] J. Austin, E. Smith, S. Srinivasan, F. Sánchez, Social Dynamics of Gang Involvement: A Mathematical Approach. Technical Report MTBI-08-08M, Arizona State University, 2011, http://mtbi.asu.edu/research/archive/paper/social-dynamics-gang-involvement-mathematical-approach.

[2] J. Boaler, *What's Math Got to Do with It?: Helping Children Learn to Love Their Most Hated Subject — And Why It's Important for America.* New York: Penguin, 2008.

[3] C. Boyd, A. Casto, N. M. Crisosto, A. M. Evangelista, C. Castillo-Chavez, C. M. Kribs-Zaleta, A Socially Transmitted Disease: Teacher Qualifications and Dropout Rates. Technical Report BU-1526-M, Cornell University, 2000, http://mtbi.asu.edu/research/archive/paper/socially-transmitted-disease-teacher-qualifications-and-high-school-drop-out-.

[4] F. Brauer C. Castillo-Chavez, *Mathematical Models in Population Biology and Epidemiology.* New York: Springer, Second Edition, 2012.

[5] D. Burkow, C. Duron, K. Heal, A. Vargas, L. Melara, A Mathematical Model of the Emission and Optimal Control of Photochemical Smog. Technical Report MTBI-08-07M, Arizona State University, 2011, http://mtbi.asu.edu/research/archive/paper/mathematical-model-emission-and-optimal-control-photochemical-smog.

[6] E. T. Camacho, C. M. Kribs-Zaleta, S. Wirkus, The Mathematical and Theoretical Biology Institute — A Model of Mentorship Through Research, *Math. Biosci. Eng. (MBE)*, **10** (5/6) (2013) 1351–1363.

[7] C. Castillo-Chavez, C. W. Castillo-Garsow, Increasing Minority Representation in the Mathematical Sciences: Good Models But No Will to Scale Up Their Impact. In R.G Ehrenberg, C.V. Kuh, eds., *Graduate Education and the Faculty of the Future.* Ithaca, NY: Cornell University Press, 2009.

[8] C. W. Castillo-Garsow, Teaching the Verhulst Model: A Teaching Experiment in Covariational Reasoning and Exponential Growth. PhD thesis, Arizona State University, Tempe, AZ, 2010.

[9] C. W. Castillo-Garsow, Continuous Quantitative Reasoning, in R. Mayes, R. Bonillia, L.L. Hatfield, S. Belbase, eds., *Quantitative Reasoning and Mathematical Modeling: A Driver for STEM Integrated Education and Teaching in Context. WISDOMe Monographs*, Vol. 2. Laramie, WY: University of Wyoming Press, 2012.

[10] C. W. Castillo-Garsow, The Role of Multiple Modeling Perspectives in Students' Learning of Exponential Growth, *Math. Biosci. Eng. (MBE)*, **10** (5/6) (2013) 1437–1453.

[11] C. W. Castillo-Garsow, C. Castillo-Chavez, S. Woodley, A Preliminary Theoretical Analysis of An REU's Community Model, *PRIMUS: Problems, Resources, Issues Maths. Undergraduate Stud.*, **23** (9) (2013) 860–880.

[12] C. W. Castillo-Garsow, H. L. Johnson, K. C. Moore, Chunky and Smooth Images of Change, *For Learning Maths. (FLM)*, **33** (3) (2013) 31–37.

[13] L. Catron, A. La Forgia, D. Padilla, R. Castro, K. Rios-Soto, B. Song, Immigration Laws and Immigrant Health: Modeling the Spread of Tuberculosis in Arizona. Technical Report MTBI-07-06M, Arizona State University, 2010 http://mtbi.asu.edu/research/archive/paper/immigration-laws-and-immigrant-health-modeling-spread-tuberculosis-arizona.

[14] K. Chow, X. Wang, R. Curtiss, C. Castillo-Chavez, Evaluating the Efficacy of Antimicrobial Cycling Programmes and Patient Isolation on Dual Resistance in Hospitals, *J. Biol. Dynam.*, **5** (1) (2011) 27–43.

[15] G. Chowell, A. Cintron-Arias, S. Del Valle, F. Sanchez, B. Song, J. M. Hyman, C. Castillo-Chavez, Homeland Security and the Deliberate Release of Biological Agents, in A. Gumel, C. Castillo-Chavez, D.P. Clemence, R.E. Mickens, eds., *Modeling the Dynamics of Human Diseases: Emerging Paradigms and Challenges.* American Mathematical Society, 2006, pp. 51–71.

[16] G. Chowell, P. W. Fenimore, M. A. Castillo-Garsow, C. Castillo-Chavez, Sars Outbreaks in Ontario, Hong Kong and Singapore: The Role of Diagnosis and Isolation as a Control Mechanism, *J. Theoret. Biol.*, **224** (2003) 1–8.

[17] N. M. Crisosto, C. M. Kribs-Zaleta, C. Castillo-Chavez, S. Wirkus, Community Resilience in Collaborative Learning, *Discrete Contin. Dynam. Syst. Ser. B*, **14** (1) (2010) 17–40.

[18] D. Daugherty, J. Urea, T. Roque, S. Wirkus, Models of Negatively Damped Harmonic Oscillators: The Case of Bipolar Disorder. Technical

Report BU-1613-M, Cornell University, 2002, http://mtbi.asu.edu/research/archive/paper/models-negatively-damped-harmonic-oscillators-case-bipolar-disorder.

[19] S. Del Valle, A. Morales Evangelista, M. C. Velasco, C. M. Kribs-Zaleta, S. F. Hsu Schmitz, Effects of Education, Vaccination and Treatment on HIV Transmission in Homosexuals with Genetic Heterogeneity, *J. Math. Biosci. Eng.,*, **187** (2004) 111–133.

[20] K. Diaz, C. Fett, G. Torres-Garcia, N. M. Crisosto, The Effects of Student-Teacher Ratio and Interactions on Student/Teacher Performance in High School Scenarios. Technical Report BU-1645-M, Cornell University, 2003, http://mtbi.asu.edu/research/archive/paper/effects-student-teacher-ratio-and-interactions-studentteacher-performance-hig.

[21] J. L. Dillon, N. Baeza, M. C. Ruales, B. Song, A Mathematical Model of Depression in Young Women as a Function of the Pressure to be "beautiful". Technical Report BU-1616-M, Cornell University, 2002, http://mtbi.asu.edu/research/archive/paper/mathematical-model-depression-young-women-function-pressure-be-beautiful.

[22] A. M. Evangelista, A. R. Ortiz, K. Rios-Soto, A. Urdapilleta, U.S.A. the Fast Food Nation: Obesity as an Epidemic. Technical Report MTBI-01-3M, Arizona State University, 2004, http://mtbi.asu.edu/research/archive/paper/usa-fast-food-nation-obesity-epidemic.

[23] B. Flyvbjerg, Five Misunderstandings About Case-Study Research, *Qualitative Inquiry*, **12** (2) (2006) 219–245.

[24] J. Gjorgjieva, K. Smith G. Chowell, F. Sanchez, J. Snyder C. Castillo-Chavez, The Role of Vaccination in the Control of Sars, *J. Math. Biosci. Eng.*, **2** (4) (2005) 753–769.

[25] B. González, E. Huerta-Sánchez, A. Ortiz-Nieves, T. Vázquez-Alvarez, C. Kribs-Zaleta, Am I Too Fat? Bulimia as an Epidemic, *J. Math. Psychol.*, **1** (47) (2003) 515–526.

[26] M. Herrera-Valdez, M. Cruz-Aponte, C. Castillo-Chavez, Multiple Outbreaks for the Same Pandemic: Local Transportation and Social Distancing Explain the Different Waves of A-H1N1PDM Cases Observed in Mexico During 2009, *Math. Biosci. Eng. (MBE)*, **8** (1) (2011).

[27] I. Kareva, F. Berezovskaya, C. Castillo-Chavez, Myeloid Cells in Tumour–Immune Interactions, *J. Biol. Dynam.*, **4** (4) (2010) 315–327.

[28] C. Kribs-Zaleta, M. Lee, C. Román, S. Wiley, C.M. Hernández-Suárez, The Effect of the HIV/AIDS Epidemic on Africa's Truck Drivers, *J. Math. Biosci. Eng.*, **2** (4) (2005) 771–788.

[29] G. Leinhardt, O. Zaslavsky, M. K. Stein, Functions, Graphs, and Graphing: Tasks, Learning, and Teaching, *Rev. Educational Res.*, **60** (1) (1990) 1–64.

[30] R. May, Simple Mathematical Models with Very Complicated Dynamics, *Nature*, **261** (5560) (1976) 459–467.

[31] B. Morin, L. Medina-Rios, E. T. Camacho, C. Castillo-Chavez, Static Behavioral Effects on Gonorrhea Transmission Dynamics in a MSM Population, *J. Theoret. Biol.*, (2010).

[32] A. R. Ortiz, D. Murillo, F. Sanchez, C. M. Kribs-Zaleta, Preventing Crack Babies: Different Approaches of Prevention. Technical Report BU-1623-M, Cornell University, 2002, http://mtbi.asu.edu/research/archive/paper/preventing-crack-babies-different-approaches-prevention.

[33] O. Prosper, O. Saucedo, D. Thompson, G. Torres-Garcia, X. Wang, C. Castillo-Chavez, Modeling Control Strategies for Concurrent Epidemics of Seasonal and Pandemic H1N1 Influenza, *Math. Biosci. Eng. (MBE)*, **8** (1) (2011) 141–170.

[34] M. Resnick, *Turtles, Termites, and Traffic Jams: Explorations in Massively Parallel Microworlds.* Cambridge, MA: The MIT Press, 1997.

[35] K. Reusser, Problem Solving Beyond the Logic of Things. Textual and Contextual Effects on Understanding and Solving Word Problems, in *70th Annual Meeting of the American Educational Research Association*, San Francisco, CA, April 1986.

[36] K. R. Rios-Soto, C. Castillo-Chavez, M. Neubert, E. S. Titi, A. Yakubu, Epidemic Spread in Populations at Demographic Equilibrium, in A. Gumel, C. Castillo-Chavez, D. P. Clemence, R.E. Mickens, eds. *Modeling the Dynamics of Human Diseases: Emerging Paradigms and Challenges*, American Mathematical Society, 2006, pp. 297–310.

[37] K. R. Rios-Soto, C. Castillo-Chavez, B. Song, Epidemic Spread of Influenza Viruses: The Impact of Transient Populations on Disease Dynamics, *Math. Biosci. Eng. (MBE)*, **8** (1) (2011) 199–222.

[38] D. M. Romero, C. M. Kribs-Zaleta, A. Mubayi, C. Orbe, An Epidemiological Approach to the Spread of Political Third Parties, 2009, arXiv:0911.2388.

[39] W. Rudin, *Real and Complex Analysis.* New York: McGraw-Hill, Third Edition, 1987.

[40] F. Sanchez, M. Engman, L. Harrington, C. Castillo-Chavez, Models for Dengue Transmission and Control. in: Modeling the Dynamics of Human Diseases: Emerging Paradigms Challenges. in A. Gumel C. Castillo-Chavez, D.P. Clemence, R.E. Mickens, eds. *Modeling the Dynamics of Human Diseases: Emerging Paradigms and Challenges.* American Mathematical Society, 2006, pp. 311–326.

[41] A. H. Schoenfeld. *On Mathematics as Sense-Making: An Informal Attack on the Unfortunate Divorce of Formal and Informal Mathematics*, Hillsdale, NJ: Lawrence Erlbaum Associates, 1991, pp. 311–343.

[42] S. Seal, W. Z. Rayfield, C. Ballard II, H. Tran, C. M. Kribs-Zaleta, E. Diaz. A Dynamical Interpretation of the Three-Strikes Law. Technical Report MTBI-04-07M, Arizona State University, 2007, http://mtbi.asu.edu/research/archive/paper/dynamical-interpretation-three-strikes-law.

[43] B. Song, M. A. Castillo-Garsow, K. Rios-Soto, M. Mejran, L. Henso, C. Castillo-Chavez, Raves, Clubs and Ecstasy: The Impact of Peer Pressure, *Math. Biosci. Eng. (MBE)*, **3** (1) (2006) 249–266.

[44] P. W. Thompson, Conceptual Analysis of Mathematical Ideas: Some Spadework at the Foundation of Mathematics Education, in O. Figueras, J. L. Cortina, S. Alatorre, T. Rojano, A. Sepulveda, eds. *Proceedings of the Annual*

Meeting of the International Group for the Psychology of Mathematics Education, Vol. 1. Mexico: Morelia, 2008, pp. 45–64.

[45] P. W. Thompson, In the Absence of Meaning, in K. Leatham (ed.) *Vital Directions for Mathematics Education Research.* New York: Springer 2013.

[46] A. Yakubu, R. Saenz, J. Stein, L. E. Jones, Monarch Butterfly Spatially Discrete Advection Model, *J. Math. Biosci. Eng.*, **190** (2004) 183–202.

Chapter 8

Integrating Mathematics Majors into the Scientific Life of the Country

William Yslas Vélez

Department of Mathematics, University of Arizona
Tucson, AZ 85721, USA

Research Experiences for Undergraduates (REU) programs in mathematics are a wise national investment, an investment that is designed to produce the next generation of mathematicians. Even though these programs target undergraduates, their impact has been much broader. They have changed mathematical culture. As just one important example of cultural change: thirty years ago it was rare for undergraduate mathematics majors to conduct research, now it is widespread, and even considered an important factor in applications to graduate schools and prestigious fellowship programs. When we look abroad we see that other countries have not made this change.

REU programs are designed to promote post-graduate study in the subject area. Participating in an REU provides much more than just a research experience for a student. Among the benefits to undergraduates are:

- Open-ended problems are presented, with no clear solution. As opposed to coursework, students are given the opportunity to explore, to conjecture and to formulate plans of attack.

MSC 2010: 01A65, 01A72 (Primary); 01A80 (Secondary)

- Computing is often a large part of an REU program and students either learn a programming language or apply their computing skills.
- Group work is often included in REUs and this activity creates a network of friends and contacts. An important research tool is emphasized: talking to each other.
- Career guidance is provided at REUs and students are given information about graduate schools.
- Faculty and graduate students who run the REU program can serve as a resource for students as they proceed along their educational path.
- The existence of REU programs serves as motivation. It shows students that hard work can lead to exciting paid travel opportunities early in their careers.

The impact of REUs goes beyond the undergraduates and has reached into faculty ranks and mathematical culture in the following ways:

- Faculty who run REUs have accepted the challenge of developing research problems for students with minimal mathematical background.
- The REU environment is not the classroom and the faculty/student interaction is different.
- Graduate students often participate in REU projects. This participation can be viewed as professional development for the graduate students. These graduate students could become the next generation of REU leaders.
- The interaction between faculty and graduate students can be more collegial in an REU setting.
- Undergraduates are now presenting their results at the national conferences and undergraduate participation is increasing at these conferences.

But, who can participate in these REU programs?

Mathematicians create questions or problems for students to work on in REUs. This is quite a daunting task and mathematicians should be commended for their efforts to engage undergraduates in this fashion. The fact that students are to be engaged in research mandates that they arrive with some minimal knowledge. Most REU programs are for students who have completed the first two years of the mathematics major course of study, plus some more advanced coursework. This requirement gives the impression that third-year students would have an opportunity to apply to these REUs, that is, REUs are open to potentially all math majors at least once during their undergraduate careers.

This is far from the truth. Most mathematics majors do not select the mathematics major upon arriving at the university. Many add the mathematics major during the second or third year of study. These students most likely will not have the above-mentioned characteristics even at the end of their third year of study, so these students do not see themselves as competitive for these REU programs. Students who attend universities with small numbers of mathematics majors also have the problem that they cannot take the required courses by the end of their junior year in order to be competitive for REUs.

How do students even know that they should apply to REU programs? Getting the information about research opportunities to students, at a time that they can utilize that information, is a challenge.

How can we provide a research experience to capable students that would serve to encourage them to pursue post-graduate studies in mathematics-based careers?

There are a few REU programs that are aimed at a different audience, namely students who are just completing their first or second year of study. If REU programs are designed to encourage students to pursue graduate studies in the mathematical sciences, then such a program is a riskier investment. Nevertheless, such programs can show students a different view of mathematics, and this alone can be motivating to students. Motivating students to the continued study of mathematics is certainly a goal that mathematicians would support. However, there are simply too many students and too few REU programs to accommodate all of these students. We would all agree that providing a research experience for undergraduate mathematics majors is a sound educational goal. It is apparent that the classroom setting will not provide such a research experience and that some structured out-of-class activity is in order. Such activities take up faculty time and this is also a barrier to providing this important experience to a large percentage of the undergraduate mathematics majors.

So, how are we to provide research experiences to first and second year mathematics majors? We recognize that these students, early in their studies, may not want to proceed onto graduate studies in the mathematical sciences, or even complete the mathematics major! Nevertheless, providing these students with a research experience so early in their studies can be transformative.

There is another reality that we must confront. In stating that we want to encourage students to pursue advanced degrees in the mathematical sciences, we have to understand what that statement means.

1. What Does Graduate Study in the Mathematical Sciences Mean?

In the middle of the last century, graduate study in mathematics began with courses in abstract algebra, real/complex analysis, and topology. Mathematics departments, run by research mathematicians, structured the undergraduate program of study with a view towards graduate study and the upper division coursework of mathematics majors included courses that would prepare mathematics majors for this transition into graduate school. The upper division courses consisted of linear algebra, abstract algebra, complex variables, advanced calculus, plus other courses that represented the research interests of faculty.

Of course, there are other career paths that a mathematics major could pursue, like teaching mathematics in high school. But even for this career path, the above-mentioned, upper-division courses play a prominent role, as the following quote shows, "Begle was dealing with high school teachers who are traditionally required to complete the equivalent of a major in mathematics. However, the requirements for math majors are designed mainly to enable them to succeed as mathematics graduate students, ..." [8].

This traditional mathematical preparation of mathematics majors is still very much a part of the culture of mathematics departments. Epsilon-delta (advanced calculus) and structure (abstract and linear algebra) are considered essential ingredients in the course work for an undergraduate mathematics major, even though graduate programs in the mathematical sciences have changed since the 1950s. This same mathematical preparation may now be serving to discourage students from pursuing post-graduate studies in the mathematical sciences.

In the latter part of the last century there began a growth in graduate programs in applied mathematics. Many of these programs have connections to the life sciences, engineering, finance, among other areas. These are still mathematics graduate programs but their emphasis may be in applications. It is more appropriate now to talk about graduate programs in the mathematical sciences. Mathematical sciences includes traditional programs in mathematics, but it also includes applied mathematics, statistics and biostatistics. The last three do not require the traditional training of the mathematics major. Graduate programs in applied mathematics would like to see students with courses in differential equations, numerical analysis, linear algebra and advanced calculus. Graduate programs in statistics require advanced calculus, linear algebra, probability and statistics. Graduate programs in biostatistics state that they require three semesters of calculus and linear algebra. Notice that none of these programs require abstract algebra.

As a number theorist I greatly value the study of abstract algebra, but my responsibility in preparing the next generation of mathematical scientists is not to mold them to my career, but rather to prepare them to participate in the mathematical enterprise of this country.

2. The Importance of Options for Mathematics Majors

Attracting students into the mathematics major is but the first step, retaining students in the major is an issue that must also be confronted. The structure of the undergraduate program of study is an important ingredient in the ability of a department in attracting and retaining mathematics majors. The traditional program of study for the mathematics major has one goal in mind. Prepare students for advanced study in mathematics.

The first two years of the mathematics major are fairly typical: the calculus sequence, differential equations, perhaps a lower level course in linear algebra or some course in discrete mathematics, and some course helping students to transition into "proof". This last course is necessary as it is common for mathematics majors to take the same course in calculus as engineering students. This broad audience for calculus results in less of an emphasis on proof and students have to be provided some guidance and transition into this important tool.

There is a tremendous change in the sophistication in the mathematics major courses from the lower division courses to the upper division courses, and this change sometimes comes as a surprise to students as they transition into abstraction. Students who performed well in calculus and differential equations, and therefore select mathematics as a major, might struggle with advanced calculus and abstract algebra. One way of supporting these students is to have options for the upper division courses.

As an example of developing options I will describe the program of study at the University of Arizona (UA).

3. The Program of Study at The University of Arizona

The philosophy of the mathematics major program at the University of Arizona is that by the time students complete the undergraduate degree in mathematics, students will have functioned as mathematicians, that is

- Mathematics majors have studied mathematics and have taken at least one year-long course in some topic of mathematics
- Mathematics majors have done research or applied their mathematical knowledge in some area outside of mathematics
- Mathematics majors have participated in some activity where they communicated mathematical ideas

In addition to these points, mathematics majors belong to a mathematical/scientific organization, perhaps the undergraduate math club, and they have resumes. They are professionals.

Mathematics majors at the UA are required to take at least one course in computer programming. In fact, the more programming experience a mathematics major has, the more employment opportunities there are for that student. A minor is required for all mathematics majors at the UA and these minors are an important component in finding opportunities for research and employment. My advice to mathematics majors who wish to find employment upon completion of the bachelor's degree is that their minor should be computer science. To have the problem solving skills of a mathematician, combined with the ability to utilize the power of computing, makes for wonderful employment opportunities. The department has made it easier for students to double major in mathematics and computer science. Several mathematics courses can be used to satisfy upper-division computer science requirements in the computer science major and one course in computer science can be used to satisfy an upper-division mathematics requirement. Students who graduate with double majors in mathematics and computer science are much sought after. These majors are hired by Microsoft, Google, Lockheed-Martin, etc.

The first two years of the mathematics major at the UA are fairly typical: the three semester calculus sequence, linear algebra, differential equations, and an introduction to analysis. The above-mentioned courses are taken by all mathematics majors, no matter their career track. After this core, students need to select one of seven options, each option requires five or six more mathematics courses. One of these options is the pre-college teaching option and students planning on becoming middle or high school mathematics teachers earn their degree from the mathematics department.

Three of the options are described below.

- The Comprehensive Option is the traditional one: semester-long courses in linear algebra and complex variables and year-long courses in advanced calculus and abstract algebra.

- The General/Applied Option consists of five courses: methods in applied mathematics, mathematical modeling, matrix theory, and a year-long course in either differential equations, probability/statistics or numerical analysis.
- The Probability/Statistics Option consists of five courses: advanced calculus, probability theory, statistics, stochastic processes, linear algebra.

A very important point must be made about the undergraduate mathematics major program at the UA. The goal of the mathematics major program is not to produce students pursuing graduate programs in the mathematical sciences. The goal of the program is for students to take enough mathematics that they are enabled to reach the goals that they have set for themselves. Many of our students have double majors or are working on two degrees. This is very difficult to do in four years. It sometimes happens that students will change the mathematics major to a mathematics minor. This is not considered a failure. These students are pursuing other options having taken more mathematics than is required for their major.

Since 2010, each year there have been over 600 mathematics majors (more than 20% of these majors are minority students) and another 700 mathematics minors. Each year since 2010 we have graduated over 100 mathematics majors. About half of the graduates had another major. About a quarter of the graduates pursued post-graduate study, but most post-graduate areas have not been in the mathematical sciences. Many of these students selected the General/Applied Option and this extra mathematics made them more competitive for graduate programs outside of mathematics. These students were accepted into graduate programs at prestigious universities like Stanford, Harvard, MIT, Oxford, and Berkeley.

In 2011, the American Mathematical Society (AMS) recognized the Math Center in the Department of Mathematics at the UA with an award for an exemplary program in a mathematics department [5, 6]. In the next section the activities of the Math Center will be described.

4. The Activities of the Math Center in Increasing the Number of Mathematics Majors and in Increasing Participation in Research

The Math Center was established more than twenty years ago. The Math Center is dedicated exclusively towards running the math major program, providing advising for students, maintaining a data base, keeping track of activities for mathematics majors, sending out weekly messages to mathematics majors on opportunities and developments, and keeping in touch with the faculty advisors. Almost all the mathematics faculty have between ten and twenty advisees.

The Math Center has a full time staff person as Coordinator of the Math Center. A faculty member, currently W. Y. Vélez, serves as Director of the Math Center.

The academic year begins with a four-hour Orientation workshop for potential or new mathematics majors on the Saturday before classes start. We stress the following points.

- Students can select one of seven options available to the mathematics major. Each option is geared towards different career paths.
- Undergraduate mathematics majors who have had study-abroad experiences, summer internships or participated in research program give brief descriptions of their experiences.
- The importance of having an up-to-date resume is presented and students are emailed a "Sample Resume" on which to construct their own.
- In an attempt to form communities, students who are taking the same course are encouraged to form study groups. Emails are exchanged among students.
- The undergraduate mathematics club presents its activities, including free tutoring for selected courses.
- Employment opportunities as Undergraduate Teaching Assistantships in the department are described.
- Research opportunities in other departments are described and the importance of taking a programming course as early as possible is stressed.

We want all of our students to apply for extra-curricular experiences, including research and teaching. As we all know grades can be a barrier. Students need to know early on that their grades will be their calling cards for years to come. But there are two other major impediments.

Every application requires one-three letters of recommendation. Where are students going to obtain the support of faculty to apply for these positions if they do not attend office hours? The Orientation workshop stresses the importance of attending office hours and establishing contact with faculty so that faculty can support the students' goals.

The other major impediment is that students often view themselves as unqualified for research positions, even when they satisfy the minimum requirements. Students should understand that their role is to apply. It is not also to serve as evaluator of their own application. If they have the minimum qualifications, they should apply.

5. Integrating Mathematics Majors into the Scientific Life of the University

We want first-year students to apply for research positions. However, these students are for the most part not competitive for national REU programs. Since we have suggested to these students to take a programming course in their first year, students will find that there is some demand for students with computing skills. But where to find meaningful scientific employment?

There is a strong tradition in the life sciences to have undergraduates work in labs. Moreover, it is common for first and second year students to be hired in these labs. Given the fact that there are increasing applications of mathematics to the life sciences, I encourage first-year mathematics majors to take a biology or chemistry course in their first year.

Let's look at the program of study of a typical, first-year mathematics major. By the end of the first year, these students will have taken a year of biology or chemistry, a programming course, and two semesters of calculus. If a mathematics major applies for a position in a biology lab, it is likely that the mathematics major looks better prepared than a typical biology major. Moreover, there are many mathematics majors who arrive with credit for first semester calculus, so by the end of the first year of study, these students may be in third semester calculus and/or linear algebra. Mathematically, these students look like second-year students.

But it gets better. There are dry labs on campus, that is, there are life science researchers whose research is more computational in nature or the results of their experiments are large amounts of data. Some of these researchers would prefer to have student workers whose background includes more mathematics and programming skills. Our mathematics majors are perfect for these labs!

I had a phone call from the director of a research program asking me to recommend mathematics majors for the research project. I asked what the research was about and he said that they were investigating water quality issues. I replied that it appeared to me that biology or chemistry majors would be more appropriate. The director said that this work required the use of mathematical models and it was his experience that "it was easier to teach the science to a mathematics major than the mathematics to a science major."

My experience has been that well-prepared mathematics majors, with supporting courses in programming and science, can find meaningful research experiences outside of the mathematical enterprise. These research experiences can sometimes divert the academic goals of the mathematics major. Students may become more interested in the research area of the project than in continuing on towards advanced degrees in the mathematical sciences, or even in the mathematics major. I consider this to be a success. Our role as advisors to undergraduate students is not to force a particular path on them, but rather to help students discover for themselves where their interests and passions lie. Moreover, these students bring with them a more substantial mathematical background to the new subject area. Hopefully students will see the need to pursue more mathematical training as they proceed along their chosen academic careers. This infusion of mathematically trained students into other subject areas is only a plus.

In order to provide advice to mathematics majors about the opportunities that exist on campus for mathematics majors, the department has to be knowledgeable about their existence. Many academic departments have a long history of undergraduates working on research projects with faculty and the university most likely has a record of these activities. These opportunities are often announced on-line and on the departmental websites. National grants involving undergraduates in research are common and the university can provide information about them. Faculty in the mathematics department can be working with collaborators in other departments and this can provide further resources. The department could begin collecting information about research opportunities in other departments which can be made available to the advisors.

There is nothing more powerful than simply calling up researchers and volunteering to provide good candidates for their undergraduate research program. One could ask, "What would you like to see in a student? What background, what courses would make a student attractive to your research group?" By having this conversation with faculty in other departments the mathematician would learn better how to advise students.

One of our goals, as advisors to our students, should be to integrate undergraduate mathematics majors into the scientific life of this country. If mathematics majors could have a research experience outside of mathematics in their first or second year of study, followed by research experiences in the mathematical sciences, we would produce more well-rounded graduates, some of whom might choose to pursue further mathematical studies, but with a better understanding of the way that mathematics is applied.

There are other research or internship opportunities available for students outside of the university setting. Industry and national labs have programs that support students both in summer and during the academic year. Most of these programs will require programming and mathematics majors with programming skills can be competitive. A department could make contact with local industry to see what they would like to see in a student and this could inform the department about the proposed course of study for mathematics majors. Career fairs are common at universities and the department could send a representative to the career fairs to discuss with the recruiters what they would like to see in a student.

This brings to mind the importance of having mathematics majors create resumes. Resumes are a common feature for students in engineering and business, but not so common for mathematics majors. Why? Perhaps one of the reasons for this is that the common view of mathematicians is that mathematics majors are headed towards two standard careers, pursuing graduate studies in the mathematical sciences or becoming pre-college mathematics teachers. In fact, the majority of mathematics majors do not follow either of these two career paths. Mathematics majors pursue jobs in a wide variety of business areas or pursue graduate studies outside of the mathematical sciences.

The importance of mathematical training in other academic areas has changed dramatically over the last few decades. Students who wish to pursue graduate studies in economics would be well served to have had courses in advanced calculus, probability, statistics, and linear algebra, essentially a mathematics major at our university. Mathematical modeling is now pervasive in the life sciences and senior-level courses in differential equations, linear algebra, probability and statistics make students more competitive. Even in graduate programs in engineering, the above-mentioned courses increase the competitiveness of the student applicant. These facts bring a new dimension to both the question of finding opportunities for research opportunities for mathematics majors and increasing the number of mathematics majors.

6. The Unique Role that Mathematics Holds in University Studies — an Outreach Activity

Mathematics holds a unique position in university studies. Suppose that X is some good major that a student selects. If the student adds the mathematics major to her program of study, then X and mathematics is great, no matter what X is. The study of mathematics gives the student analytical and problem solving skills. Even if the student joins the workforce and does not do mathematics, those skills remain. Of course a person in X could argue that X also builds analytical skills and provides problem solving abilities, and this is certainly true. However, the study of mathematics provides something else. Tools! We provide the ideas and tools that are used in modeling, and the ability to use these tools is a hard-won skill, one that starts in calculus, proceeds on to a more in-depth study behind the limiting processes, then onto the study of structure. Along the way, the mathematics major is developing the ability to deal with precision and to understand when a theorem is applicable by looking at the assumptions of that theorem. In an age when many calculations are done with packaged software, a mathematics major knows when to question the appropriateness of the use of that software for a problem at hand. Yes, X and mathematics is great and students with this combination are more competitive on the job market or in graduate school.

The realization that X and mathematics is a great combination led me to develop an outreach program in my department. In March of each year, I obtain the names and email addresses of all students who have been accepted into our university. The data also includes the major that the student has selected. I have crafted letters for all of the different majors and the departmental staff sends out about 15,000 messages in March and April. We use software that takes the Excel spreadsheet, takes the name of the student and inserts the name as, Dear ***, so it appears like a personal message. The sender of the email appears to be me. So if a student replies to the email, the reply will come to me.

The tone of the letter is one of concern for the welfare of the student. For example, it contains a link to the scholarship website of the university. The letter first of all congratulates the student for having been accepted to the university. I go on to comment that the major that the student has chosen is a good one and if it is appropriate, I provide a link to where the student can look for research opportunities on campus for that major. For each letter I also provide a link to the AMS website, "Mathematical Moments", where the link is to some description of the application of mathematics to X, the student's chosen major. The Mathematical Moments articles are just a few paragraphs. I then go on to suggest that since mathematics is so critical to X, the student should consider adding mathematics as another major because X and mathematics make a great combination. I also provide a link to the math major website. This letter suggests that the student enroll in the highest level mathematics course that they are prepared for when they arrive at the university. They should not postpone the mathematics course necessary to their major.

I also provide a hook at the end of the letter. In a postscript I tell the students that if they plan to enroll in calculus that they should send me a message and I will reply with a copy of "Resources for calculus students". The two-page document describes the lower division mathematics courses, provides a link to the departmental calculus webpage and gives some advice. For example, a question that I am often asked is whether or not the student should enroll in calculus II if the high school student has earned college credit for calculus I through Advanced Placement credit. This is an impossible question to answer. The advice given on this document is that if the student is thinking about doing this, then the student should go to the calculus webpage. There the student can find the final exam questions for calculus I that have been collected over the last few years. The student should work on these problems over the summer. If the student can do most of the problems, then the student is ready for calculus II, otherwise the student should start in calculus I.

There is another good reason for placing that hook at the end of the letter. If the student replies then it gives me a chance to provide some guidance or advice to the student before the student has even arrived at the university. When the student replies I send the Resource document and I also send along a link to some videos that we created, which I hope will motivate students to add the mathematics major. I also send along a four-year schedule of mathematics courses that shows a sample schedule for the mathematics major.

Here are some examples of replies from students.

- Thank you for the helpful information. I watched the videos and they made me even more excited for college. The video about the excelling high school student who did not pass her first math test and was dropped down a level made me realize that this summer instead of sleeping and watching tv, I will take initiative and brush up on my Pre-Calc and Calc skills. Thank you for sending me those videos. They were very interesting and raised my awareness that college is definitely not like High School. College is much harder. I learned from those videos that if I continue working hard and use my resources, I will do well. Thank you for your time.

- I have always been interested in mathematics, and after looking through the online handbook, I see that there is a sensible connection between life sciences and mathematics. I am interested in learning more about the necessary steps that are required to join a degree in physiology with a degree in mathematics.
- Thank you so much for taking the time to type such a detailed response. The email is very helpful in cementing my decision. Firstly, I would love to be at the Orientation on the Saturday before classes. I think that will be a great opportunity to find out more information. Thank you for the resources on how to prepare. I want to be refreshed on the concepts I have learned already, so I will start reviewing now. I watched the videos and read the article by Sean Howe. Those gave me confidence in adding on a math minor. The benefits to more math education seem numerous, one of which, as you mentioned, would be an advantage in the workforce.
- Thank you again for the insight and replying so quickly. I found those videos to be interesting as well as eye-opening. I particularly liked the video, "Incoming Math Students", it proved to be very helpful. I now understand the responsibility and initiative you need to be prepared to have at the UofA. I am going to take full advantage of my resources now to stay ahead. Looking over the attached document, I would have never thought to start completing assignments as well as reviewing beforehand. I found that to be very knowledgeable. I can't thank you enough for your time and help. I look forward to my experiences with Math at the UofA.

In reading over these comments, and the comments of many other students, I am struck by the fact that most students do not understand the usefulness of mathematics in their careers. Most high school students have probably never heard of a mathematics major, except for those who plan to teach in high school. To me this represents a challenge to the mathematics community. How do we convey to the high school community the importance of mathematical training and the many career opportunities there are for those who are mathematically trained?

One of my objectives in sending these messages is to have students understand the huge transition that they are about to encounter as they enter college. Students just can't envision for themselves this dramatic change and many students are not prepared for it. I want students to prepare for this change.

A message from a professor inviting the student to add the mathematics majors can be very powerful. Over and over again I receive messages from students telling me how surprised and pleased they were to get a message and to be offered an opportunity to sit with a professor for twenty minutes to talk about their academic careers.

If the data that we have on the students includes the email of a parent or guardian then we copy that first message to that person. Parents can be quite impressed when a professor writes a personal note to their child. I have had several messages from parents expressing their gratitude. Parents get the feeling that someone cares about the career of their child, even in a university with 40,000 students.

Here are a couple of responses from parents.

- Thank you so much for your email and sharing with us the idea of H. considering an additional degree in mathematics. M., H. and I will certainly discuss your email. We are looking forward to our registration trip in June. I'm sure they will have a full schedule for us but if possible, and if you are there, we will come by and introduce ourselves. It was very thoughtful of you to reach out to H.
- I wanted to take a short moment to thank you for reaching out to N. and I over the last summer to introduce us to the opportunity to take Math 223 (Calculus III) as well as a dual major that included

> Math. N. is extremely excited about what he has learned over the last semester and is truly looking forward to continuing his efforts in Math 215 (linear algebra) in the fall. Your guidance was spot on and N. is excited (as am I) about his continued experiences within the math department.

I should point out that N. will be graduating in May 2015 with a major in biochemistry and minors in mathematics and computer science. Without the initial message he probably would not have considered taking so much mathematics. With the added minors he will be more competitive as he looks for post-graduate opportunities.

The messages that I send out not only increases the number of mathematics majors, but it also increases the number of mathematics majors who also have another major. Approximately one-third of all mathematics majors at the UA have another major or are working on another degree in our department. It is impressive how many students are willing to work this hard to accomplish this. Since these students are double majors, these students can apply to REU programs in their other major and this greatly increases the number of opportunities available to them. In fact, we have many students who arrive with two semesters of calculus and Advanced Placement credit for some science courses. If they decide to apply to REU programs in X during their first year of study, these students actually look like second- or third-year students in X, which again makes them even more competitive.

In summary, I would like to collect together these ideas into some bullet points.

- Mathematics holds a unique place in the educational enterprise. By encouraging students to include more mathematics in their undergraduate curriculum, we are opening up opportunities for them. Having students add mathematics as a major or minor further increases the career paths that they can pursue.
- Establishing an office in the mathematics department which focuses exclusively on the mathematics majors will help a department not only manage its math major program, but also it serves as a focal point for undergraduate activities.

- Graduate programs in mathematics have morphed to graduate programs in the mathematical sciences. Preparation for these graduate programs is more diverse than it used to be and the preparation of mathematics majors should reflect this.
- Mathematics majors, with a background in programming and some science, have many internship and research opportunities, both on campus and in industry.
- The department could take an active role in locating research and internship opportunities for its majors. Making connections to other departments on campus serves as good public relations for the mathematics department.

References

[1] Undergraduate Math Major Website at the University of Arizona, http://math.arizona.edu/ugprogram/.

[2] Sample Resume and Career Assistance, http://math.arizona.edu/ugprogram/mcenter/careerassistance.html.

[3] Videos and other resources, http://math.arizona.edu/ugprogram/mcenter/resources.

[4] Mathematical Moments, http://www.ams.org/samplings/mathmoments/mathmoments.

[5] AMS 2011 Award for an Exemplary Program or Achievement in a Mathematics Department, *Notices Amer. Math. Soc.*, **58** (5) (2011) 718–721, http://www.ams.org/notices/201105/rtx110500713p.pdf.

[6] Arizona Math Center Wins AMS Award, *Notices Amer. Math. Soc.*, **58,** (5) (2011) 716–717, http://www.ams.org/notices/201105/rtx110500718p.pdf.

[7] Advising as an Aggressive Activity, *FOCUS, Newsletter Math. Assoc. America,* **14** (4) (1994) 10–12, http://math.arizona.edu/~velez/AdvisingAggressive.pdf

[8] Minority Calculus Advising Program, http://math.arizona.edu/~velez/MinorityCalculusAdvisingProgram.pdf.

[9] H. Wu, The Mis-education of Mathematics Teachers, *Notices Amer. Math. Soc.*, **58** (3) (2011) 372–384.

Chapter 9

The Gemstone Honors Program: Maximizing Learning Through Team-based Interdisciplinary Research

Frank J. Coale, Kristan Skendall, Leah Kreimer Tobin and Vickie Hill

Gemstone Honors Program, Honors College, University of Maryland

0100 Ellicott Hall, College Park, MD 20742, USA

gems@umd.edu

Introduction

The Gemstone Honors Program resides within the Honors College at the University of Maryland as a unique multidisciplinary four-year research program for selected undergraduate Honors students of all majors. Under guidance of faculty mentors and the Gemstone staff, teams of students design, direct and conduct significant research, often exploring the interdependence of science and technology with society. Gemstone students are members of a living-learning community comprised of fellow students, faculty and staff who work together to enrich the undergraduate experience. This community challenges and supports the students in the development of their research, teamwork, communication and leadership skills. In the fourth year, each team of students presents their research in the form of a thesis to experts, and the students complete the program

MSC 2010: 97D40, 97D40 (Primary);

with a Gemstone Honors Citation, a published thesis and a tangible sense of accomplishment.

The Gemstone Honors Program was founded in 1996 in the School of Engineering at the University of Maryland. This novel initiative was launched in response to feedback from employers of Engineering students who were very impressed with the technical skills and individual capabilities of recently hired graduates but, alternatively, were discouraged with the new employees' abilities to function in technically diverse groups and effectively contribute to multidisciplinary team efforts. The original bold mission that was crafted to guide the Gemstone Honors Program in 1996 continues to motivate achievement toward four time-tested goals: (1) develop students' research skills in the context of multidisciplinary team research projects; (2) develop students' ability to work effectively in teams; (3) provide students with leadership opportunities through peer mentoring, teaching and community service; and (4) provide students with a close-knit community that supports them in all of their commitments and activities at the University of Maryland.

The Gemstone Honors Program is committed to the holistic development of scholars, citizens and leaders, with efforts focused both inside and outside of the classroom. Through fostering intellectual excitement, collaboration, and diversity of thought, Gemstone students achieve transferable skills that will be valuable in all future endeavors.

Program Structure

Today, the Gemstone Honors Program student body is composed of a total of approximately 550 Honors College students spanning the freshman through senior years. Gemstone students typically represent approximately 50 different academic majors within the University of Maryland. While the majority of Gemstone students major in STEM disciplines, the wide array of Gemstone student majors include accounting, aerospace engineering, anthropology, biochemistry, biology, computer science, criminal justice, economics, education, English, environmental science, journalism, history, kinesiology, mechanical engineering, music, mathematics, neurobiology, physics, psychology, and many more. Each incoming class cohort of approximately 175 students spends the entire first year developing and focusing their research interest and forming interdisciplinary teams of 8 to 14 students that design, direct and conduct significant original research. Typically, 11 to 13 teams are formed per class cohort, for a total of approximately 36 teams actively conducting research over the sophomore, junior and senior years. Each Gemstone research team is guided and supported by a dedicated faculty mentor with expertise in the team's selected field of research.

The Gemstone professional staff manages the Program and creates the structure that permits and fosters student attainment of the Program's goals. The Gemstone Program Director is a tenured faculty member with commitment to Program administration, development, operations and management. The Gemstone Associate Director is a full-time professional who oversees the academic functions, team research, and support of the faculty mentors and librarians who work with the Gemstone teams. The Gemstone Assistant Director for Student Engagement coordinates the student development, leadership, co-curricular and residential components of the Gemstone Program. The Gemstone Assistant Director for Operations and Team Support oversees the financial, logistical, facilities, scheduling and student records activities. Two staff coordinators support the operations of the Assistant Directors.

The Gemstone Honors Program is a priority of the University of Maryland. Approximately 80% of the Program's total operating budget is directly allocated from the University's central administration, 10% is derived from competitive sources within the University and 10% is garnered from sources external to the University of Maryland. Individual Gemstone student research teams are encouraged to pursue, and are supported in their efforts to procure, specific research project funding from a wide array of grant programs, both internal and external to the University. However, procurement of extramural funding for support of specific research teams is unpredictable, unreliable and inconsistent. The University's annual financial commitment to the Program is essential and fundamentally determines the size of the Program, student enrollment and sophistication of the research endeavors.

Admissions, Program Selection and the First-Year Experience

All students who submit a complete application to the University of Maryland by the priority deadline and are admitted to the University are subsequently considered for admission into the Honors College. The Honors College admissions team conducts a whole-file application review and admission is based, in part, on academic achievement (high school GPA), rigor of high school curriculum, standardized test scores (SAT or ACT), extracurricular activities, school community involvement and demonstration of leadership. There are no established numerical thresholds for admission to the Honors College. Following admission to the Honors College, incoming students indicate their preference for placement in any of the Honors College's seven living and learning programs, one of which is Gemstone, in which they hope to participate. Practically all of the newly admitted Honors College students who indicate Gemstone as their first preference are enrolled in the Gemstone Honors Program. Incoming students who rank the Gemstone Honors Program as their second or third preference out of the seven Honors College program options may be invited to join Gemstone after consultation and placement evaluation by the program directors. It is rare for an incoming Honors College student who expressed keen interest in participating in the Gemstone Honors Program to be denied admission.

The Gemstone Honors Program offers an invigorating and engaging first-year experience. Newly admitted freshman students arrive to campus three days prior to the arrival of the general student body to participate in Gems Camp, an overnight orientation program designed to introduce new students to the Program, create connections with their peers and provide a forum for focused guidance and advice from upper-class students serving as "Camp Leaders". In small groups, incoming students spend three days at an off-campus campsite becoming familiar with the Gemstone curriculum, leadership opportunities and University of Maryland resources, in the context of a fun and purposeful camp theme. Students leave Gems Camp feeling more confident about their upcoming Gemstone experience, more comfortable with the University and energetic about embarking on their four-year research journey with many new friends.

The Gemstone Curriculum

The Gemstone Honors Program employs a formal and sequential curriculum to guide the formation of the undergraduate research teams and development of a rigorous research process, which is capstoned with a published thesis. To complete their Gemstone Honors Citation, all students in the Gemstone Honors Program are required to take all of the Gemstone courses offered, which amounts to a total of 18 credits over four years. The courses in the first year are designed to orient the students to the University of Maryland and the Gemstone Honors Program and to serve as an introduction to research conducted in teams and across disciplines. In the first semester, students take GEMS100 — Freshman Honors Colloquium: Introduction to Gemstone, which is a one-credit course taught by Gemstone staff designed primarily to assist with the students' transition to college. In the second semester, freshman students enroll in two Gemstone staff-taught courses which total four credits and are designed to be complementary courses as students explore potential research topics and form research teams. During GEMS102 — Research Topic Exploration, as student propose and evaluate potential team research topics and team memberships, they participate in a series of ice-breaking and intentional team building activities that evolve into consensus-based decision-making activities and structured problem-solving exercises. GEMS102 students also participate in improvisational activities to help advance their listening and collaborating skills. Concurrently, students are enrolled in GEMS104 — Topics in Science, Technology, and Society, which is a fast-paced, team-oriented, writing-intensive, one-semester microcosm pre-exposure to the upcoming three-year Gemstone research experience.

In the fall semester of the sophomore year, all students enroll in GEMS202 — Team Dynamics and Research Methodology, a two-credit course taught by the Gemstone professional staff and designed to help the teams create and write their Gemstone thesis proposal. In GEMS202, students also participate as teams in a high-ropes challenge course that has proven to be very conducive to promoting intra-team bonding and communication. Concurrently, beginning in the sophomore year, students also take the first of six project research seminars (GEMS296), which is where they begin formulating, planning and executing the actual research project. Gemstone students enroll in project research seminars during every semester and, over six semesters, the project seminars account for 11 of the 18 required Gemstone academic credits. The project seminars are the times when teams meet with their faculty mentor and work to achieve established milestones and the faculty mentor is the instructor of record for these research seminar courses. In GEMS296, fall semester of the sophomore year, and GEMS297, spring semester of the sophomore year, students finalize their research proposal, initiate research activities, begin the process of applying for grants for both internal and external funding support and acquire research protocol approvals, such as Institutional Review Board (IRB) or Institutional Animal Care and Use Committee (IACUC). Frequently, a subset of the student team will continue research activities through the summer months following the sophomore year, either as volunteers or as employees paid by the faculty mentor's research group.

In the junior year, Gemstone students enroll in GEMS 396 in the fall semester and GEMS 397 in the spring semester. The junior year is focused on data collection and analysis. Students are required to present their preliminary findings in the fall semester as an oral presentation at the Gemstone Junior Colloquia and again in the spring semester as a poster presentation at the University of Maryland Undergraduate Research Day. Again, it is common that a subset of the student team will continue research activities through the summer months.

In the senior year, students enroll in project seminars (GEMS496, fall semester; GEMS497, spring semester) that focus their time and energies on completing their research and writing the final team thesis. In the spring semester of the senior year, Gemstone seniors present their final thesis to a committee of experts at the Gemstone Senior Thesis Conference. In addition to earning their Gemstone Honors Citation upon completion of the program, Gemstone students also fulfill many of the University of Maryland's general education requirements, conduct original research, build close and lasting relationships with faculty mentors and publish a team thesis.

Research Team Formation

Gemstone student research teams are formed during the spring semester of freshman year during the GEMS102 — Research Topic Exploration course. Students propose research concepts individually or in small groups and present the potential research topics to their peers. The preliminary research concept prospectuses include a problem statement, a research question, background on the topic, and names of possible faculty mentors. Also, the Gemstone Honors Program Director annually solicits potential research topics from all faculty and senior staff on the University of Maryland campus. Faculty-proposed projects are presented to the freshman students along with peer student-proposed projects. Historically, very few research topics have been proposed by mathematics faculty. Most naïve freshman students have difficulty envisioning the applicability of math-centric research aims to multidisciplinary exploration of the interdependence of science and technology with society. It is incumbent upon the mathematics faculty to propose such concepts in a context that intrigues and excites the freshman students.

After presentation of all of the project concepts to the entire cohort, the students vote for their five most favored projects and the pool of possible research topics is narrowed based on popular opinion. After elimination of unpopular research ideas, a topical expert must vet each of the remaining project concepts for feasibility, scope, and relevance. Following the vetting process, a second round of student voting occurs and the list of potential and viable research ideas is narrowed to approximately 14 topics. At team formation, the final vote is conducted by having each student specify his or her top three team topic choices. Students are guaranteed placement onto one of their top three choice teams and, through an intentional process that includes preferences for research project concept initiators, student participation in the process, and guidance from upper-classmen section leaders who have worked closely with the students, the freshman students are placed onto their final Gemstone research teams.

Gemstone research team topics range from tackling questions in the physical and biological sciences and engineering to challenges in the behavioral and social sciences to inquiries in the humanities. Individual Gemstone research teams are purposely multidisciplinary and are composed of students with a wide variety of academic majors and scholarly interests. Over the years, the Gemstone Honors Program has purposefully aimed to diversify the backgrounds, interests and academic goals of the Gemstone student body. Melding diverse students together on a single Gemstone team has greatly contributed to positive intra-team dynamics, expanded cross-discipline learning and ultimate team success.

Although, historically, teams have not formed around purely mathematical research questions, elements of applied mathematics have been incorporated into the majority of Gemstone undergraduate research projects at both the team and individual levels through implementation of tools such as MATLAB or quantitative modeling. For example, Gemstone Team BIOCOUNTER (Bioweapon Inhibition Operating Containment Unit for the Negation of Terrorist Entities and Radicals) completed their research in 2013 with a project that used both qualitative and quantitative methods to better understand the spread of and response to a potential anthrax attack in the Washington, DC, metropolitan area. Following interviews with government officials, Team BIOCOUNTER used mathematical modeling to measure the efficiency of anthrax spore detection devices and the theoretical time-responsiveness of designated responding agencies utilizing plume dispersion models to simulate the spread and detection of anthrax spores. This undergraduate research team used mathematical modeling as a tool to understand ways in which various interventions could be optimally deployed in an effort to minimize the human harm created by a theoretical anthrax attack. Individually, many Gemstone students study mathematics as a major course of study, minor, or as part of an engineering or other science major. Because Gemstone research teams are interdisciplinary, these students not only apply their knowledge on Gemstone teams, they also share that knowledge with their fellow teammates. Some students take on particular aspects of a project as their own and apply their mathematics knowledge to that part of the project. One current example is a student on Gemstone Team BASS who is a mathematics major and is utilizing mathematical modeling to better understand the relationship between rainfall patterns, animal manure application to agricultural fields, water quality toxicological data, and the incidence of ill-defined fish gender or fish hermaphroditism, which is a tangential aim to the team's overall project goals.

Faculty Mentors

First and foremost, Gemstone team research projects are formulated and developed by Gemstone students. After a team of freshman students has crafted a project concept, the Gemstone staff begins the process of mentor selection from the University of Maryland faculty and senior professional staff. All faculty mentors are volunteers who have expertise in the field of study selected by the Gemstone team and professional interest in fostering undergraduate research. Faculty mentors commit to serving in a supporting, guiding and facilitating role to the Gemstone team for three years. A modest salary supplement is provided by the Gemstone Honors Program for each year the faculty member serves as a Gemstone mentor. In return, formally, the mentor must agree to meet with the student team for at least one hour per week during the academic semester and support and facilitate team progress. In reality, most team mentors are deeply dedicated to their student teams and allocate more than the required minimum time commitment. Faculty mentors assist the student teams with navigating a variety of challenges related to lab safety and technique training, necessary laboratory protocol approvals, funding, equipment, and space. Most team mentors share their lab space and equipment with their teams or assist the Gemstone staff with identifying and securing appropriate research facilities and equipment. The faculty mentors work with the students to develop the skill-sets necessary to define concise research questions, propose suitable methodological procedures, conduct the research and evaluate the results. Frequently, mentors solicit support from professional colleagues to assist with elements of the research experience that would benefit from collaborative effort. Identification of eager and dedicated discipline-appropriate faculty mentors is a crucial element for ensuring the success of the undergraduate research teams. Broadly, the University academic culture must support and encourage faculty to embrace serving as a Gemstone mentor as a positive professional opportunity and not view this mentoring role as an imposed burden. There must be pervasive mutual recognition of the educational value of the Program throughout all facets of the University, from undergraduate students to the faculty to the offices of senior administration.

Some University faculty members have served as the mentor for multiple Gemstone teams over many years. Returning faculty mentors are testament to the reciprocal benefits derived by mentors from engaging with the talented and energetic Gemstone Honors students. Rarely, however, have University faculty with academic roots in mathematics served as mentors for Gemstone undergraduate research teams.

Team Librarians

The uniqueness of the Gemstone Honors Program, the quality of its students, and the Program's tremendous success brings great credit to the University of Maryland. The Program's success is partially due to partnerships built with the University Libraries. Annually, dedicated librarians support 36 Gemstone research teams through their literature review, proposal development and thesis preparation. The librarians dedicate their time to assist Gemstone teams and offer direction and guidance on developing background research and literature synthesis skills. Individual librarians are assigned to assist between one and three research teams and remain in the supportive instructional role with the team throughout the duration of the team's research project. The librarians are topical experts with degree backgrounds ranging from the humanities and social sciences to the biological and physical sciences and engineering. Their technical knowledge on current issues and experience with academic research have been essential elements contributing to team success. The Gemstone Honors Program staff hosts librarian workshops and training sessions each year to orient new librarians to the goals and processes of the Program and to acquaint experienced librarians with new team research topics. The librarians are very active in critiquing presentations at the Junior Colloquia and poster presentations at Undergraduate Research Day. Gemstone librarians are honored participants at the annual Gemstone Thesis Conference and attend the Gemstone Citation Ceremony to celebrate with their graduating senior teams. The librarians play an instrumental role in the development of teams' research skills and the Gemstone teams benefit from the close working relationships developed with the professional librarians.

Co-curricular, Leadership and Service Activities

The Gemstone Honors Program prides itself on not only offering a unique four-year research experience for talented undergraduate students, but also on providing many out-of-class leadership opportunities. Students may serve as a "Section Leader", which is a role similar to that of an undergraduate teaching assistant, for several courses (GEMS100, GEMS102, GEMS202 and GEMS104). In these roles, Section Leaders help guide the students through the course material and provide valuable Gemstone upper-class experience throughout the different phases of the research planning and methodology development processes. Students may opt to receive either a monetary stipend or class credit for their service.

The Gemstone Student Council (GSC) offers additional leadership opportunities students may pursue within the Program. The Gemstone Student Council is comprised of an executive board (President,

Secretary, Treasurer, Vice President for Academic Affairs, Vice President for Community Engagement, Vice President for Student Activities) and three larger student committees led by each Vice President. All Gemstone students are welcome to attend regular committee meetings and participate in the regular GSC general body meetings. The GSC provides social and educational programming to support the Gemstone student body and serves as a representative voice for the student population as a direct connection to the staff.

Gemstone students may also participate as Gemstone CONNECT mentors, which are students who serve as peer mentors and facilitate and host evening study sessions in the Gemstone office suite several evenings a week. Gemstone CONNECT student mentors make themselves available to provide support to fellow Gemstone students needing assistance with course work, provide general college living advice, and maintain and supervise a quiet place to study. Students may also participate on the Gemstone Outreach Team, which is a group of students who volunteer to assist with University Orientation sessions during the summer, open house days for prospective or admitted students, Visit Maryland Days and many more events. At all of these events, enthusiastic Gemstone Outreach Team students regale visitors, recruits and families about the impressive Gemstone story.

Lastly, Gemstone students may serve on planning committees for GEMS 100 and Gems Camp. These positions allow seasoned Section Leaders and Camp Leaders to participate in the planning process for course syllabus revisions and upcoming events. Both planning teams are student-led and staff-advised, to allow returning Gemstone students to have the highest level of input into the design and content of both the GEMS 100 course and Gems Camp. Again, students may opt to receive either a monetary stipend or class credit for their service.

Program Success

The Gemstone Honors Program is successful in its goals of fostering the holistic development of scholars, citizens and leaders through stimulating intellectual excitement, collaboration, and diversity of thought. Although participation in the Gemstone Honors Program requires tremendous dedication to the students' research project, peer team and faculty mentor; Gemstone students are intently involved in campus life and extend their education beyond our campus. Over 90% of Gemstone students were members of a campus organization beyond Gemstone and half of those students have held leadership positions in their chosen campus organizations. Seventy percent of Gemstone students held an internship at some point in their college careers. One-third of Gemstone students have had a study abroad experience and executed a specific team-learning agreement prior to studying abroad that serves as a contract between the students who are abroad and their teammates. This agreement outlines the contributions to team progress that will be delivered by the globetrotting students during their absence from campus and typically includes a schedule for regular live video communication with the team during team meetings.

By and large, Gemstone students overwhelmingly acknowledge satisfaction with their Gemstone experience, enjoy the sense of community the Gemstone Program creates and are proud of their Gemstone experience. As of 2013, the three-year mean four-year graduation rate for Gemstone students was 84%, compared to 67% for the University of Maryland as a whole, and the six-year graduation rate for Gemstone students was 94%, compared to 83% for the University (personal communication, Office of Institutional Research, Planning and Assessment, University of Maryland, 2014). In recent years, approximately 40% of the Gemstone graduates transitioned directly to professional employment, 20% pursued graduate degrees in a wide array of disciplines, 20% attended medical school, 5% attended law school, and the remaining 15% engaged in pursuit of other advanced professional degree programs and internships.

Summary

The Gemstone Honors Program, a vibrant component of the Honors College experience at the University of Maryland, is truly unique.

This multidisciplinary four-year research program allows teams of students to design, direct and conduct original research while being active members of a stimulating living-learning community. Since the inception of the Gemstone Honors Program in 1996, the declared majors of Gemstone students have increasingly represented the vast array of academic disciplines available across our campus, including mathematics. However, student-derived team project topics focusing on questions in mathematics have been rare, if existent. Also, voluntary participation by mathematicians through submission of faculty-proposed research concepts has been equally rare. Thus, the involvement of mathematicians as Gemstone faculty mentors has lagged behind the involvement of faculty from other STEM disciplines, apparently due to constraints within the mathematics discipline or from cultural traditions among the mathematics faculty. Math faculty participation is not limited by the structure of the Gemstone Honors Program. The Gemstone Honors Program maximizes undergraduate learning inside, outside, alongside and beyond the campus classroom. The model employed by Gemstone may be readily applicable to the revitalization of undergraduate mathematics learning. It is incumbent upon faculty mathematicians to think broadly and envision mathematics research concepts that may be presented to and adopted by multidisciplinary teams of freshmen undergraduate students that are committed to exploring the interdependence of science and technology with society.

Chapter 10

The Freshman Research Initiative as a Model for Addressing Shortages and Disparities in STEM Engagement

Josh T. Beckham, Sarah L. Simmons and Gwendolyn M. Stovall

Freshman Research Initiative, College of Natural Sciences
The University of Texas at Austin
Austin, TX, USA

James Farre

Department of Mathematics, College of Natural Sciences
The University of Texas at Austin
Austin, TX, USA

Problem Statement/Motivation. In the United Sates, there currently exists a shortage of STEM (Science Technology Engineering and Math) graduates from colleges and universities [1]. Of those that do graduate with these degrees, they are often underprepared for the rigors of the workplace due to the majority of their didactic experience coming from lecture-based formats [2, 3]. Higher education administrations are beginning to see the need for change in the way that science is taught to more effectively foster students' abilities to assimilate, create and present scientific knowledge.

MSC 2010: 97D40, 97D40 (Primary); 01A73 (Secondary)

Addressing the change in diversity which is bringing lower projected graduation rates. There is even a more pressing need for change in states such as Texas due to the rapidly changing demographic makeup. In 2010, Hispanic population increased in 229 of the 254 counties in Texas [4]. While this shift is beneficial in providing diversity to the state, it also brings with it, if we extrapolate current trends, lower graduation rates across disciplines. These projected disparities in educational attainment extend into the sciences in particular. The involvement of minorities in bachelor-level work is reflected through the matriculation numbers at The University of Texas at Austin. As of 2011, the entering new undergraduate population at UT Austin-consisted of 23.6% Hispanic, 5.2% African-American and 0.3% American-Indian [5]. In contrast, the State of Texas' demographic was 37.6% Hispanic, 12.6% African-American and 1.3% American-Indian [4]. These discrepancies reflect that fewer minorities are entering the state's largest university and have disproportionately low involvement in the sciences relative to their demographics' size in the population. Considering that high paying jobs necessitate college degrees, these individuals will be at a disadvantage. Yet, the projections are not foregone conclusions. If widespread interventions are made to direct the younger generation of Texans into colleges and through their degree plans, they will not have to be subject to these statistics. Rather, they may be able to obtain skilled positions and contribute to maintaining the country's economic advantage [6].

Research engagement as a solution to the STEM shortage. It is well documented that research can help retain students in college and within the STEM disciplines [2, 7–9]. Research projects which engage students in real world, applied work also develops their critical thinking skills and content knowledge. We also know that early research experience profoundly affects graduate school performance and long term academic and career success [7]. Steps towards curriculum reform of undergraduate education with these end goals in mind is advancing nationally [2, 6]. Institutions are beginning to understand that to involve significant numbers of students, elements of research and the scientific method need to be built into the curriculum so all students are impacted and not just those students who would have gotten involved in research based upon their backgrounds or aptitudes.

In response to the need for comprehensive change, The University of Texas at Austin (UT-Austin) has developed an innovative model, the Freshman Research Initiative (FRI), to ensure that a significant number of undergraduates have the opportunity to do authentic research within the College of Natural Sciences (CNS). The program addresses two fundamentals: engaging students in applied research and doing so at an early time point. However, unique to the FRI program is that the engagement is on a large enough scale to be significant relative to the shortages in adequately trained STEM graduates. The program was started in 2005 as a pilot with 43 honors students that carried out research under a faculty member as part of course credit towards their majors in chemistry, biochemistry and molecular biology. These groups were called "research streams" and each were instructed by a non-tenure track faculty that worked with the primary faculty member. The following year, the program was able to expand to four times its size and incorporate non-honors students through an NSF grant (CHE #0629136) and also an HHMI (Howard Hughes Medical Institute) grant (#52005907, 52006958). The program is now at 25 research streams across many more disciplines (biology, chemistry, biochemistry, computer science, physics, astronomy, and math) and takes in approximately 800 new students each year. The program is predominantly maintained through HHMI funds and internal funds from the College of Natural Sciences' instructional budget. Individual principal investigators also contribute by allocating some of their grant funds to the research. Often this support is part of the broader impacts of a funded proposal.

The model harnesses the twin missions of a research institution of education and research, in a mutually beneficial way. The teaching aspect is improved through the use of research as a learning environment, while the research aspect benefits through the acceleration or broadening of the science. Many large universities already effectively take advantage of this synergy by providing traditional one-on-one, mentor–student relationships in faculty research labs.

However, within large institutions these opportunities affect too few students for too short a time period at a point that is too late in their education. Additionally, recent reductions in research funding and the current faculty reward structure discourage the use of resources for undergraduates in research. The FRI model has been able to overcome some of these institutional challenges and incorporate large numbers of undergraduate students into original research activities at their home institution when they are still taking lower-division credits. The University of Texas at Austin, with over 40,000 undergraduates, is one of the largest state universities in the country and is among the top public research universities in the world. The scholarship within College of Natural Sciences (CNS) itself is reflected by the $110 million in annually sponsored research to over 700 tenured and non-tenure track faculty members [10]. CNS itself houses 11,200 undergraduates and 1370 graduates within 11 departments. In this tenth year of the FRI program, approximately 800 students were enrolled into the spring lab experiences. Over the 9 years since the inception, more than 4500 undergraduates have been impacted by the program.

Course Sequence

Overview. Students who participate in the FRI are guided through a process over their first year and a half which has the potential to be extended over successive semesters [11]. They are recruited during orientation as incoming freshman students to begin the program in their first semester (fall), through enrollment in a Research Methods course. In the second freshman semester (spring), they enroll in one of our research "streams", combining training in research skills which are the beginning of their authentic research projects. Students can continue over the optional summer experience or choose to continue in the fall of their sophomore year, for academic credit. In the second semester of their sophomore year, the course sequence is complete, but some have the opportunity to serve as peer teaching assistants (known as "Mentors") to the new cohort of students entering the research stream. Thereafter, students who have participated in our program have the skills necessary to apply to be placed either with a university faculty member's lab, a research internship at a government lab, an industrial research lab, or a traditional summer research program such as an REU.

Research Methods. In their first semester as freshmen (fall), the entire cohort takes Research Methods, which is an inquiry-learning course on scientific problem solving and research methodology. Research Methods includes both a lecture and a minor laboratory component. This course provides a signature course credit (required for all UT undergraduates) to most students. Lab inquiries include freshman-level lab concepts as well as exposing students to general equipment and techniques used in research. The goal of this course is to introduce the students to the research process in general by: (a) designing small-scale research investigations, without explicit cookbook-style guidance; (b) using mathematics and statistics to summarize and model experimental findings; (c) evaluating scientific conclusions critically; (d) becoming familiar with the various forms of research, ranging from deductive to inductive, controlled environments to observations, pure to applied, and from theoretical to experimental and computational; and (e) becoming familiar with the social context of research, including research literature, research communities, funding and publication processes, and how scientific communities react to new ideas. The inquiry-learning course offers scientific problem solving and research methodology to the students, thus priming them for an authentic research experience.

Research Streams. The research stream model that begins in the spring of their first year is the heart of the FRI approach for merging the usually disparate activities of research and education. Rather than placing individual students with individual faculty, or integrating parts of research into traditional laboratory courses, the streams are fully functional research laboratories with an educational purpose. Student placement into their respective research stream in the spring is dependent upon their top five choices and, to a small degree, their performance in the Research Methods course from the fall. The stream experience involves both once a week lectures and significant research time (6–8 hours/week) in the lab. The course counts toward their degree plan in almost all cases. In the classroom, students are introduced to the methodology and scientific context for their specific research in order to engage in their own independent research projects. In the lab, the approach starts with basic techniques appropriate for otherwise inexperienced students, yet by the end of the second semester, the methods are advanced enough that students could contribute to publishable research.

In the summer, students in the stream are encouraged to stay on campus to continue work on their projects. Since many of our students remain to take standard summer courses, it is possible for them to spend time in lab as well. While the FRI summer experience is not for credit, it is a structured research program to which all of the streams participate. A limited number of summer stipends are available to the students, but several choose to continue research as volunteers simply out of enthusiasm for the program and its benefits. Often times, the summer research experience can be the most transformative as more time is spent in lab with less academic load from other courses. The summer also provides a deeper sense of community as the students bond while spending as much as 20–40 hours/week together in the pursuit of common research goals.

In the fall of sophomore year, students that choose to stay on in the program, receive an upper-division lab FRI credit as part of their stream experience. This credit most often counts in the students' degree plans. The fall course is much more open-ended and less rigid in weekly class requirements, instead focusing on research progress and a final report. Most of the student data is generated in the fall when they have been able to develop their projects. The culmination of their work is either reflected in their final reports or when they present a poster at our college-wide Undergraduate Research Forum or at other national or regional conferences. After the fall experience, students have completed their credit sequence and can choose to become mentors in the FRI, seek out research positions in other labs, or stop and take a break from research.

Personnel

Principal Investigator. Each of these research streams is led by a tenured faculty member who has designed a program of research using our educational model to pursue their fundamental research questions. The faculty PI (Principal Investigator) provides guidance to their respective research streams, setting goals and directions, and may teach part of the lecture portion of the course. A key feature of our approach is that authentic research reflecting the core interests of the faculty is performed; this ensures their involvement and commitment. Oftentimes, the PIs use the research stream as a component of their overall research wherein they may explore riskier experiments than what they would give to a graduate student who needs a publication to graduate. The power of the research from a stream can also be a function of the amount of data generated by having many undergraduates working on a project simultaneously. Furthermore, such a partnership with the FRI, often categorized as an "outreach" activity itself, offers numerous additional outreach opportunities, which are often highlighted and received favorably in NSF "Broader Impacts" sections of proposals by the PIs.

Research Educators. One of the challenges to initiating large scale undergraduate research programs is that faculty are not incentivized to be involved. Often times the considerations of publications for tenure and promotion take priority over research-based teaching. Furthermore, faculty are occupied with traditional lecture-based teaching loads and are unable to devote time and resources to overseeing many undergraduates in their labs. The FRI model circumvents these problems through the creation of the Research Educator (RE) — a novel, non-tenure track faculty position that is a hybrid of teaching and research. The RE has completed their PhD and instructs one research stream. Through the research educator, the tenured faculty member is able to take on 15–35 undergraduates in a meaningful way. The RE's main focus is the implementation of the science and the education of the student participants in the research labs. The RE oversees the stream and the day-to-day operations of the student research projects. They are the one interacting with the students in lab and teaching them the skills and procedures necessary for their research experiments in the cognitive apprenticeship model [12]. The lead faculty PI and their Research Educator work closely together by discussing strategic plans about the research aims and ensuring coordination of research and teaching. The FRI model allows flexibility for the involvement of the lead PI while retaining the key link to them that gives the stream authenticity in its work and contact with tenured faculty for the students.

The Research Educator position is also a vehicle for professional development and the transport or dissemination of the FRI model and the institutional knowledge. When these faculty move to other colleges and universities throughout the country, they will take the institutional knowledge and culture of undergraduate research with them. As opposed to a traditional postdoc where research is the exclusive priority, these Research Educators will have had direct teaching experience. The Research Educators also develop administrative capacities through organizing their class, running a research lab, managing undergraduate teaching assistants and contributing to programmatic duties of the FRI. All of these attributes make them excellent candidates for faculty positions at PUIs (primarily undergraduate institutions) or as junior faculty at larger research universities. As a sample of the Research Educators that have gone on after their time with the FRI, we have had four go on to industry jobs. One of those still collaborates with the FRI through his company. Two have gone on to traditional postdoctoral positions focusing solely on research. One RE has taken a role within the college at UT-Austin as the director for postgraduate education. Four former REs have gone on to become professors at small liberal arts institutions. Of those, one has transitioned to another institution into the role as the director and founder of a first year research program similar to the FRI.

Peer Mentors. While the Research Educator provides the primary teaching and management responsibilities for the stream experience, there still exists a need for more instructional staff to accommodate the numbers of students in a class and the demands of a laboratory environment. As with typical research labs, it is best when these individuals are the lab's existing members. In the FRI, students in the second year of the program are trained as undergraduate peer TAs (Peer Mentors) to help teach the new cohort of incoming students. Along with the ability to pass on their knowledge of the research and lab techniques, these sophomore Peer Mentors act as role models and sources of support for the new students. They have overcome the same challenges not only in research but also in college life in general. A hierarchy of the lab instruction is also supported by the presence of the traditional junior or senior level Teaching Assistants (UGTAs) or Graduate Teaching Assistants (GTAs). The maintenance of this infrastructure and communication is one of the many duties managed by each Research Educator.

The mentors not only provide organizational scaffolding that makes the stream work, but also improve the academic outcomes for the students. Student-centered learning through small group collaboration has been shown to foster higher-order thinking, deeper conceptual understanding and the development of interpersonal and leadership skills in science and math education [13–15]. Students who have previously mastered the knowledge and skills of a particular subject become effective peer teachers and can facilitate group learning [16, 17]. Evidence suggests that well-prepared peer teaching assistants may be more effective as facilitators than course instructors in keeping groups on task and providing essential strategies for successful college achievement [18–20].

We have often seen that the students who stay on as mentors benefit the most from the FRI experience. They are in a position of responsibility where their leadership skills are fostered. Once they have had the chance to teach the lab techniques to another person, they find their own understanding of course content and the research has become much more profound. Through working with a many students who have varied academic and social backgrounds, the mentors learn to explain concepts and techniques in different ways. As a result, they are more prepared for their careers where they will most likely interact with a broad demographic.

Research

The types of research implemented in the FRI program are varied and applied. Below are descriptions for several of the different research stream topics.

Aptamer Stream

The Aptamer Stream uses oligonucleotide chemistry, *in vitro* selection methodology, and biochemistry to develop novel therapeutics, diagnostics, and molecular sensors. The tool of choice for the development of these applications is an "aptamer," an oligonucleotide binding species. In the stream, students use the methods of *in vitro* selection methodology to identify aptamers against a variety of targets and develop their downstream application. For example, aptamers targeting the precursors of disease states could offer inhibitory functions, thus serving as therapeutics to alleviate disease. Still other aptamers against disease biomarkers could be used in the development of disease diagnostics. Additionally, some aptamers may serve as gene regulators (such as synthetic riboswitches) or for targeted drug delivery (such as nanoparticle adjuncts).

Nano Stream

In this research stream, students use cutting edge technology to synthesize nanoparticles using combinations of metals including copper, gold, platinum, palladium, and nickel. The particles act like "nanoreactors" to catalyze chemical reactions. Nanoparticles are tiny, around one billionth of a meter across, but they hold huge potential for use in biosensors, drug synthesis, fuel production, environmental remediation, and specialty chemical production. Our goal is to identify the best chemical catalysts that could eventually be used in these applications.

Synthesis & Biological Recognition Stream

The design of small organic molecules that bind tightly and selectively to proteins is essential for the development of potent drugs that have minimal side effects. In this multidisciplinary stream, student researchers design and synthesize novel molecules that bind to proteins and learn to express, purify, and test the proteins of interest. The strength of binding between the protein and small molecule is tested by a technique called ITC, which provides data that can be used to elucidate the chemical features that allow for stronger intermolecular interactions, contributing to the field of drug discovery.

VDS Stream (Virtual Drug Screening)

Identifying new drug leads using traditional methods is an expensive and time-consuming process. This research stream uses both computational and wet lab techniques to discover new drugs against infectious diseases. First, a molecular docking program is used to sift through libraries of chemical structures and predict which ones may bind to a protein that is a potential drug target. Results are visually analyzed with a molecular graphics program. Students then utilize Virtual Drug Screening software and molecular graphics programs to interpret the results and rank the protein–ligand binding interactions. Then DNA cloning and protein expression protocols are then implemented in the lab to test the top potential drugs in enzyme assays.

Computational Biology Stream

Research on self-replicating computer programs (digital organisms) enables students to experience evolution in action and to perform evolutionary experiments that would take years to complete with natural organisms. Digital organisms evolve to perform computational tasks. Completion of these tasks rewards the organisms with resources they can use to replicate faster and gain a competitive edge. Over time, faster-replicating organisms out-compete slower-replicating ones. Hence, the organisms evolve to complete increasingly complex tasks, in a manner that parallels the evolution of natural organisms. This stream is a good option for students who want to learn about computer science and evolutionary biology.

Autonomous Intelligent Robotics Stream

The goal of this computer science stream is to create a system of fully autonomous robots inside the new Gates–Dell complex that can coordinate with each other to aid people inside the building. Students will learn about and contribute to cutting-edge research in artificial intelligence and robotics.

Electronic and Magnetic Materials Stream

This stream focuses on material physics and the design and development of new materials with electronic and magnetic properties for use in data storage, optics, sensors, and optical and infrared astronomy. Students learn new materials synthesis via solid-state reaction, followed by structural, microscopic, magnetic, thermal, and superconducting characterization of the materials. Students may also perform cantilever micromagnetometry, interferometry, and magnetic resonance microscopy.

Space. Streams are housed within dedicated lab space that may or may not be attached to the PI's area. Weekly lectures or group meetings are held in typical classrooms on campus or within the labs themselves. Having dedicated lab or meeting space enables a greater sense of community where the students not only conduct research together, but can study and socialize. Ideally, the lab spaces are structured to provide open communication between the instructional staff (RE and Mentors) and also amongst the student researchers. For example, vertical shelf space in the middle of the lab is minimized to allow better line-of-sight interactions and discussion. In contrast to graduate labs, the FRI lab spaces often trade storage and bench space in favor of capacity.

Benefits & Outcomes. Over the last 9 years of the program, we have been able to observe student performance to garner a better understanding of the benefits and outcomes of the FRI experience (Data in preparation). First, student demand for the program has been increasing each year. The FRI incoming class has grown from ~45 students to over 800. The waitlist of students that are seeking a spot has grown to ~150 in this past year. This growth reflects support from the administration, enthusiasm from students, and an implied endorsement from their parents who are particularly influential in their first semester course selections.

The FRI program has been shown to successfully attract students in risk populations. One of the key benefits that we have seen through this program is the ability of research to "level the playing field" for underrepresented groups. The playing field is "leveled" by actively recruiting "at risk" groups and through the course-based research approach. Research provides a space to succeed for these students even though they may have poor academic legacies from their high school or family. Initial successes in research are often about effort and tenacity over content knowledge and intellect. The College of Natural Sciences has identified five groups that have traditionally shown difficulty in completing their degrees within the college: (1) Underrepresented Ethnicities (American-Indian, African-American, Hispanic), (2) First Generations students (first in their family to go to college), (3) Women in Target Majors (Astronomy, CS, Physics, Math), (4)

Low SAT Scores (<1100), (5) Low Socioeconomic Status (<$40k of family income). Through active recruitment to these individuals and their families upon their matriculation to the university, the program has been able to match the representation of entering students relative to that of the College as a whole for the groups of: Underrepresented Ethnicities, Women in Target Majors, and Low Socioeconomic Status. However, the groups of Low SAT and First Generation students are still underrepresented in our program relative to the population of the college. Yet a profound impact is being made in recruitment since in 2011, the FRI cohort consisted of over 50% of its students coming from one of these underrepresented groups.

In addition to a rise in demand, the tracking data has also shown improved outcomes for the students that have been through the FRI [21]. In order to make valid comparisons, a control cohort was selected. This control group consisted of students who were on equal academic footing to FRI students but chose not to be involved in the program and rather seek their degree requirement through traditional courses. Key to this selection was the Predicted GPA (grade point average) which the University of Texas at Austin uses to categorize incoming freshmen based upon their prior academic record and achievement in high school. The FRI group and Control Cohort we chose so as to have equivalent predicted GPAs. Also, to limit the effect of self-selection, both the control cohort and the FRI cohort consisted of students that had expressed interest or awareness of research when filling out an orientation survey. Honors students are excluded from the groups in order to exclude a bias toward research and a higher rate for successful completion of natural science degrees. As a result of these conditions, the comparison groups are demographically similar.

The FRI program yields more science graduates. Retention within the sciences shows that more FRI students stay within the College of Natural Sciences and graduate with a natural sciences degree than their counterparts. The effect upon retention is even more pronounced for Hispanic students within the FRI program.

After graduation, more FRI students go on to graduate or professional schools than the control group. While this metric does not directly monitor their success in these pursuits, the data implies that FRI students direct themselves towards further education.

The FRI program has not only been able to benefit the undergraduates themselves, but has contributed to the greater scientific community. A goal of each stream is peer-reviewed publications and presentation at conferences through oral talks and poster presentations. To this end, FRI students have coauthored over 100 papers that have either been published or are in press. Many of these students may then go on to contribute to scholarly work in their respective fields after graduation using the skills and experience that they gained from their FRI experience.

The unique nature of the FRI. The success of the FRI program stems from several factors that make it unique to other programs. We believe that the structure of the program devised with the Research Educator position is a key attribute that allows the needs of scale to be met while still providing student access to PhD level instructors. The close relationships of the instructors and mentors are particularly important for fostering a welcoming learning environment for students at such an early time and at such a large institution.

The fact that students receive course credit for their research allows them to get involved without the lab time being a commitment outside of their very busy degree plans. This aspect is particularly important to reaching those students that are at risk. Having course credit attached to the research also increases its perceived value and importance to the students.

An advantage to early engagement at the freshman level is that students attain the full experience of research design, execution, presentation, writing and, sometimes, publication. Once they are able to see the big picture of the research process, they are more capable of understanding their role within a field of science. Early engagement also allows students to make a more informed decision as to whether research is something they want to pursue further. Traditionally most science majors have to wait until their junior or senior year before an undergraduate research opportunity is available. By then it is often too late for them to change their track without significant consequences in time and money.

The FRI program is also unique in that it provides this experience over a long time frame. Instead of merely taking part in research for a semester, students can be immersed for a full year. As science is a continuous process with iterative approaches to generating evidence, students can understand the scope of the work better when they are carrying out many parts of the process themselves over time [22, 23]. This continuity is also key to students developing ownership in their projects where they associated the outcomes of the project with their own contributions. While most programs offer either credit or have a strong summer program, the FRI is able to offer both to many of the students.

Students build their confidence through research. The earlier they can do this, the more impact it will have upon the rest of their academic careers. When students are immersed in a project as an individual or as a team, they take ownership. This aspect carries over into their classroom academics because they maintain a sense of responsibility for their own education similar to their responsibility to develop results in their research projects.

The cross-disciplinary nature of the FRI program is another aspect that sets it apart from other undergraduate research experiences. By supporting research projects that span multiple departments and fields of study, the FRI provides vehicles for rich academic exploration that undergraduates need to help them discover their interests and understand the scope and context of research within the sciences. In contrast, it had been noted that insular programs of just one department do not lend themselves to cross-disciplinary work since departments will often seek to protect and advance their own interests [2].

Added Community. The way in which the FRI program builds a sense of community and belonging is a particularly key factor that may contribute to its success. Estrada-Hollenbeck et al. note that belonging to a group and believing in the missions and goals of that group is crucial for positive academic and societal outcomes when students are involved in research at the undergraduate level [24]. The FRI develops this community over a full year and a half. In the first semester Research Methods course, the students are part of the program as a whole. Although it is a large group of over 800 students, they start identifying themselves as FRI students and begin to build relationships. The group has a common purpose of research. In the spring, their community is made smaller and becomes more refined as they are now identifying themselves as one class with one common research mission associated with the Principal Investigator's body of work. Over the spring the groups become closer knit as they deal with the trials and tribulations of a lab-based course together. The summer experience has shown to be one of the most profound catalysts for integrating these students into a cohesive unit. The environment is more relaxed as students can focus on their lab work without the distraction of other coursework. Students begin to associate the lab as a second home because they spend so much time there. They begin to be more involved in social events with friends in the lab. The classes often buy T-shirts with a class design on them which further brings unity. By the fall, the group is well defined and students have a support network amongst each other.

Generating Mathematics-Intensive Research Experiences

The FRI program has been very successful in engaging undergraduates in research in many disciplines, however the math-intensive research experiences have been limited. We believe that several factors contribute to this: few math students are available for such research endeavors; faculty in mathematics were not early adopters of the FRI model (described below in "MIC Stream: Leading the Work with the Math Liaison"); freshman students often have not completed the proof-based mathematics courses (e.g. discrete math, linear algebra, etc.) necessary to rigorously address questions found in math research; there is a shortage of accessible math projects for incoming freshmen; and there is a sentiment that without coursework in abstract algebra, mathematical analysis, and topology, undergraduate students lack the basic tools to attack a research-level problem in pure or applied math. With these immediate hurdles in mind, we worked to address these concerns and introduce a novel approach for supplying mathematically satisfying research experience to undergraduate students.

Mathematics in Context (MIC) Stream: Pilot Program Overview

In the spring of 2014, the pilot program, "Mathematics in Context" (MIC) Stream, was launched as a companion to the established Aptamer Stream (described above) to provide a mathematics-intensive research experience. Students enrolled in the Aptamer Stream were provided the opportunity to participate in the MIC Stream enrichment program, asking mathematical questions motivated by content they encountered through their aptamer and oligonucleotide (i.e. DNA and RNA) biochemistry research. The FRI department appointed a mathematics graduate student in the 'Math Liaison' instructional role to develop, implement, and lead this initiative. The highly functioning and highly motivated graduate student (author Farre) was instrumental in creating and launching the pilot program.

The MIC Stream incorporated the mathematical study of DNA topology and knotting problems in biological systems to further describe folding and binding phenomena observed by aptamers and other oligonucleotides. The research sought to motivate the markedly more abstract notions of modern mathematics by providing some context and motivation for its study using students' spatial intuitions, interest in molecular biology and aptamer research. A detailed reading of a foundational research article in DNA topology, "Knotting problems in Biology," by M. Delbrück (1962) provided the structure for the course [25]. Weekly class time (1 hour/week) was meant to give the groups of students, having very limited exposure to math, the tools to conduct independent inquiry and guide their focus in the direction of problems motivated by Aptamer Stream content. For example, a class time might include a brief lecture (encouraging questions) on a topic such as elementary set theory, elementary logic, metric spaces, or vector operations followed by problem solving to help students navigate through technical portions of the paper, while emphasizing the need for such language and ideas behind its development. The rest of the time in the classroom was comprised of review sessions, group discussion facilitated by the instructor, graduate student guest speakers, and/or student presentation. Students were encouraged to work on problems that were of personal interest, working in small groups, if these interests aligned. Ultimately, the direction of the 10-week course was strongly influenced by the students' interests, backgrounds, progress, and work.

MIC Stream: Drawing in the Students and Diversity

Initially, student recruitment was a concern, as few math students are in the FRI (ex. 2014 FRI cohort contained 37 math majors divided between 20+ streams). With this in mind, and due to the nature of the MIC Stream research and structure of the course (i.e. subset of established stream), we sought to engage both mathematics majors and non-majors. Therefore, recruitment efforts were designed with a "low risk, high reward" structure to encourage participation from a variety of students. For example, interested students were informed that they would gain valuable problem solving experience, the opportunity to earn a strong letter of recommendation, access to faculty in the math department through the Math Liaison, and be exposed to interesting mathematical and scientific ideas not included in traditional courses. Although the MIC Stream did not offer course credit, participation in the Stream was encouraged by providing a grade incentive in the Aptamer Stream course. Additionally, a non-binding agreement for participation was offered to students. That is, students were encouraged to participate for a trial period, before making time and work commitments. Although the MIC Stream students were still part of their parent or partnering Stream (i.e. the Aptamer Stream) and required to carry out the majority of the experiments and course assignments with the rest of the class, some of this required work could have been "replaced" with time spent on their MIC Stream work, if necessary. This arrangement was another example of how the pilot program provided a low-risk environment for students to explore mathematics, especially relevant to those on the fringes of math (i.e. students who are interested in exploring math but would not necessarily commit to being math majors so early in their career). These efforts successfully recruited six students into the MIC Stream.

One of the missions of the FRI is to recruit students from a variety of ethnic and socioeconomic backgrounds, and the MIC Stream pilot program followed suit. The MIC Stream pilot program consisted of four males and two females, and of those there were three Asian Americans, one Hispanic, one black, and one white individual. The GPA range was 2.77–3.96 with an average of 3.47 and a median value of 3.4. The students' majors were: biochemistry, undeclared, neuroscience, biology (2), and economics. They were all freshmen or sophomores with 40–75 hours in residence at the university. Two were eligible for financial aid and none of them were considered "at risk" by the university's metrics. None were first generation college students, as their parents had either college degrees or higher. The students' SAT scores ranged from 590–760 on the verbal section and 630–760 on the quantitative section of the test.

By the end of the semester, the instructor retained five of the six students who initially agreed to participate.

MIC Stream: Leading the Work with the Math Liaison

The development of a Math Liaison instructor position was one of the most fundamental necessities of the MIC Stream. Mathematics faculty were not early adopters of the FRI program, as they have been effectively disincentivized from supervising undergraduate research by pressure to publish, heavy teaching load, and graduate student research supervision. These factors are true, to varying degrees, for most faculty across disciplines. However, in mathematics there exist few successful models of integrating freshmen and sophomores into mathematics research as there have been for the natural sciences. In response, a highly functioning and highly motivated graduate student was absolutely essential to launching the MIC Stream pilot program. More specifically, bridging mathematics with research in another discipline, the development of an interdisciplinary mathematics and science educator (or a "Math Liaison") allowed us to harness the structure of the existing FRI program (i.e. similar to the previously established research educator position). The Math Liaison, knowledgeable in the relevant natural science discipline of DNA topology, formulated mathematical questions that had bearing within the partnering stream. The Math Liaison, devoting 10–20 hours per week to the MIC Stream, implemented math modules or components relevant to the overall stream research. A goal of this interface was to promote interdisciplinary dialogue and highlight a relationship between math and science wherein each discipline informs the other.

Equally important was the level of professional development achieved by the graduate student serving as the Math Liaison. The Math Liaison graduate student had direct exposure to teaching in a small group environment on an applied project as well as designing his own curricula for the various research projects. His leadership skills advanced, as he was responsible for guiding the students in their work and developing the project idea with the Aptamer Stream Research Educator. The experience permitted the Math Liaison/graduate student to lead a research project in their own area of interest, thus potentially approaching or even reaching some of their own research goals. Additionally, further professional growth was realized through research collaborations with senior faculty members, as well as through the program development with senior administrators. The experience gained through this position offered numerous advantages for future work in and outside of academia.

For later iterations of the program, we propose that a non-tenure track faculty member would be appropriate for the instructor role and could be called a "Research Educator Math Liaison". A commitment at this level would retain a talented, non-tenure track faculty member through a potentially fulfilling teaching experience wherein they could incorporate their own research and provide continuity to the students over a whole year's span. This proposal is further discussed in the "MIC Stream, v2.0: Research Educator Math Liaison" section below.

MIC Stream: Providing the Necessary Background Material for Students

The research goal of the MIC Stream was to modify Delbrück's DNA knotting mathematical model [25]. To accomplish this, the students devised their own DNA knotting model, using their knowledge of molecular biology and biochemistry. If they were able to agree on and describe their model successfully, they repeated some of Delbrück's subsequent calculations and numerical experiments.

In an effort for the research to progress while providing the students with the background material and literature reviews, students were provided early "primer" modules/introductions on the material and then immersed in the higher order problems. A flexible and adaptive curriculum was necessary to accommodate the students of varying backgrounds in mathematics. Initially, the Math Liaison highlighted the differences and similarities between research in math and science by providing context for how problems in the sciences can be given rigorous mathematical formulations. The introduction to logical statements about mathematical models and proofs was necessary to establish a framework for the work. Furthering the early work, it was necessary to introduce the language and formulation of modern mathematics, which differs from the language and problem-solving approach used in calculus and high school math. These topics were followed-up with an introduction or in-depth review of elementary set theory, logic, methods of proof, formulation of functions, additional discrete mathematics topics, metric spaces, vector operations, and modern formulation of continuous functions between topological spaces. Of particular interest were continuous, piecewise linear functions from the circle into three-dimensional Euclidean space, i.e. knots.

MIC Stream: Developing Suitable Projects

Under the MIC Stream pilot program, the mathematics-intensive research projects suitable for undergraduates were carefully crafted with a multidisciplinary approach by the Math Liaison. Taking into consideration some of the visualization cues that RNA and DNA folding and structures provide, research in DNA topology appeared most appropriate topic for study with the Aptamer Stream students that were in the MIC Stream. More specifically, students sought to address a question in knot theory, which is an area of mathematics that, historically, has lent itself to undergraduate research projects due to its pictorial presentation and concrete formulation. Using the techniques and theories obtained in the MIC Stream, the students posed a question in DNA topology based on problems relevant to the research from the Aptamer Stream.

MIC Stream: Learning Outcomes

By devising revised DNA knotting models through a mathematics-intensive research experience, the learning outcomes for the MIC Stream students were to learn some basic vocabulary (i.e. expanding mathematical literacy) and tools for speaking about mathematics in a mature, modern language. Furthermore, the Stream sought to apply this knowledge and vocabulary to create and describe a model for "knotting" RNA or ssDNA molecules.

Based on post-semester MIC Stream evaluations, conversations, and focus groups, the students gained an appreciation for how abstract mathematics can be used to solve problems in science. Additionally, students obtained hands-on experience with learning how to ask questions and which questions we can pose rigorously in a theory. As the students put it, they learned that "there is life after calculus."

MIC Stream: Pilot Program Feedback and Shortcomings

We received an overwhelmingly positive response from the MIC Stream students at the conclusion of the first semester of the pilot program. The students appreciated the small class size and the flexibility of the class. The encouragement of the students to speak freely in class gave them the confidence to provide their own solutions. Additionally, the students enjoyed the literature reading component and the fact that they had to seek out papers on their own. Likewise, the students found the course content interesting and, for the most part, even more exciting than their traditional math course. The students also felt that offering a 1-hour credit for the course would be beneficial.

We note that the instructor's initial goals for the group were somewhat ambitious and give an outline of some problems that he encountered during the pilot. The most prominent obstacle was the time constraint for in-class activities and what could reasonably be expected from students outside of class. It is clear that a two-hour weekly time commitment over ten weeks is insufficient for meaningful understanding of foundational materials. These materials are typically covered over a three credit hour discrete math course at the University of Texas. Students clearly struggled with the recall of material covered in class the week earlier, and the instructor sacrificed rigor for breadth of topics, working in such a limited timeframe.

Additionally, the Math Liaison instructor hoped to reach out to other FRI streams to try and make as broad an impact as possible. However, after connecting with students in a computer science research stream called "Autonomous Intelligent Robots," the instructor realized that his topological interests and knowledge did not transfer over to being able to direct and supervise research in machine learning. Moreover, the students in this CS stream were more mathematically mature than the students in the Aptamer stream, which were predominantly biology and biochemistry majors. It would have been difficult to cater to the skills of students in both the Aptamer and Autonomous Intelligent Robots streams in the using the same time and resources of one part-time instructor. We were therefore unable to test the viability of having the Math Liaison work with multiple streams across disciplines simultaneously. With this knowledge in mind, we can now give a revised model for implementation going forward that addresses the challenges while improving the impact upon the students.

MIC Stream, v2.0: Engaging Math Majors and Non-math Majors

Encouraged by the success of our pilot program, we intend to modify and scale-up the program to serve two populations of students: math majors and non-math majors. The first group consists of incoming freshman who have already declared a major in math and who wish to get started on research in math right away. The second group is that with which we engaged in the pilot program and is comprised of students with declared majors in another natural sciences discipline and have an interest in math—they may be on the fence about committing to a math major or minor. In both populations, there is interest in engaging with mathematical research early on. In support of multidisciplinary studies, the Bachelor of Science and Arts (BSA) was recently created by the College. This new degree encourages minors and concentrations in disciplines outside of the student's major. Students seeking this degree would have the flexibility to explore mathematics in the context of research with our revised model of implementation.

Math Majors Course Sequence. For the first cohort of students, those already majoring in math, we propose a two semester course sequence: first semester: research introduction (1 credit hour course) and standard math course (3 credit hours); and for the second semester: an upper-division research course (3 credit hours). See Figure 1 for a schematic of the course sequence. The 1-hour course in the spring would be taken in tandem with a 3-hour credit course which is a standard course in proof-based elementary linear algebra, its applications, and a number of more advanced topics in linear algebra heading toward current research-level topics. The purpose of the additional 1-hour course is to provide a venue for contextualization of the mathematics into the research problems, in part by reading and exploring math research articles. After completing this introduction to research course in the spring, these math majors would be able to take a full 3-hour credit upper-division course (M375) in the fall where they more fully engaged in addressing a research problem. The topic of the research would most likely be chosen by the Math Liaison

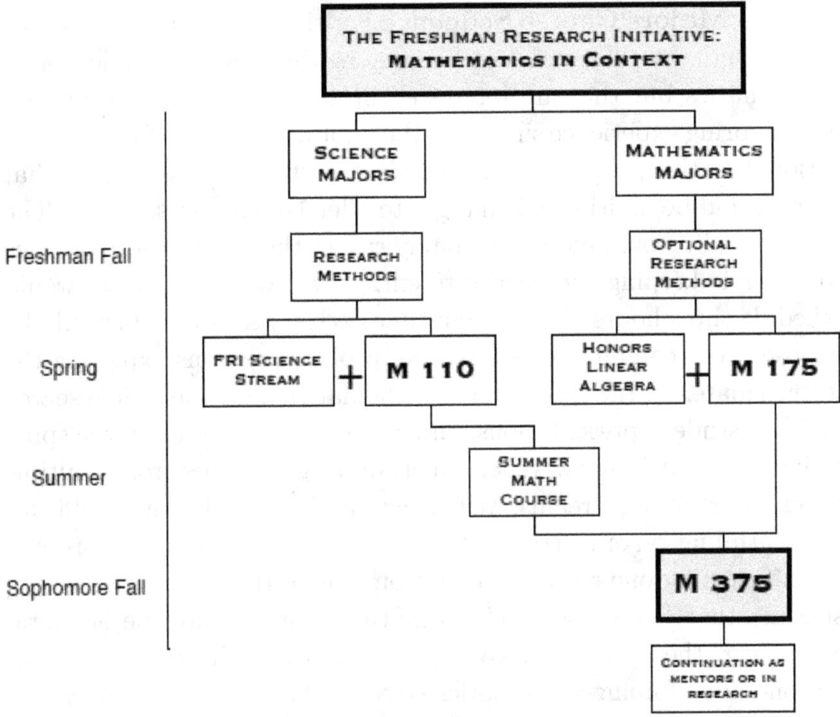

Figure 1. Course Sequence for the Mathematics in Context Stream. There are two tracks — one for science majors on the left and one for math majors on the right. Research methods introduces the concepts of research generally. Then a three credit hour FRI Science Stream or a three credit hour Honors Linear Algebra is taken along with one of the one credit hour MIC Stream specific courses (M 110 & M 175) which introduce students to topics in math research by working through a relevant research article; the mathematics is motivated through what students learn in the lab and take note of how the mathematics can inform the science. On optional course can be taken over the summer to improve math skills. In M 375, students work independently or in small groups on a mathematical research project under supervision of the math liaison. This MIC Stream specific course is the one in which students would be most engaged in the mathematics of the research.

(or instructor) to align with their strengths and subject matter of interest. By involving these students as freshmen and sophomores into research, this course sequence would provide exposure, awareness and an appreciation for the broader role of mathematics within sciences that is not normally available so early in their college careers.

Science Majors Course Sequence For the second cohort (approx. 10–15 students), natural science majors that are not majoring in mathematics but that are interested in discovering more, we would offer a spring experience similar to the pilot program. To further institutionalize the program and offer degree support, we believe that there would be multiple advantages to offer 1-hour course credit. The 1-hour credit would more adequately reflect the effort devoted to this work while keeping the barrier to entry low. Additionally, we would schedule three hours of classroom time each week. Students would be exposed to a combination of lecture, problem sessions exploring the lecture material, the reading of a mathematical and scientific research articles, student presentations, and guest lecturers — all in the spirit of independent inquiry. When incorporating students from multiple parent/partnering streams, we believe that it would be intractable for the Math Liaison or instructor to manage up to two separate topics at once. Selecting one research topic from one of the parent/partnering streams that lends itself well to math research would be advantageous, e.g. the aptamer DNA topology work or the research of the Autonomous Intelligent Robotics stream. During the pilot program, we found that students valued a collaborative approach in a less formal environment than a traditional lecture. The Math Liaison invited group discussions and deviations from prepared material to talk about the specific questions of students. Since the students would be in class for a total of three hours a week, there would not be an expectation of work done at home, other than readings.

After the spring course, the students would have the option to transition into the more rigorous 3-hour credit course in the fall by having taken a more traditional mathematics course such as linear algebra, number theory, or discrete mathematics in the summer. This standard math course in the summer would give them the skill-sets necessary to take on more challenging work of the fall 3-hour credit course and put them on a more level playing field as those in the first cohort who were math majors.

For either cohort, the math majors and the science majors, ending the program after only one semester would still be a positive outcome. For the science majors, ability to "put their toe in the water" and try out mathematics serves the broader mission of developing inter-disciplinary scientists. If they were not continuing the program because they did not enjoy mathematics, it could also be taken as a positive outcome because they could then appreciate to focus their efforts in other disciplines. For the cohort that was math majors, a one semester only experience in the MIC Stream would provide them an early window into the applications of math to scientific problems. Having the knowledge of mathematics in a broader context may then more effectively direct their future endeavors in mathematics. They would also be more scientifically literate through their time in the MIC Stream where they focused on scientific problems. Regardless of their exit point, the math majors that had participated in the MIC Stream at such an early time in their college careers would be uniquely poised to seek out further math experiences later on — for example, through REU programs within the following summers.

MIC Stream, v2.0: Research Educator Math Liaison

While the role of the Math Liaison instructor for the FRI pilot was carried out in the highly capable hands of a graduate student in Mathematics (author Farre), there would be a benefit to giving more permanence to the position. We will work towards employing a non-tenure track faculty as a Research Educator like the other streams have. The Research Educator Math Liaison would teach both courses in the spring (the 1-hour credit for majors and the 1-hour credit for non-majors) and incorporate elements of his or her own research interests into the experience throughout. They would then be responsible for summer research with the students and/or teaching the more traditional course that would be used to get the non-major students up to speed. In the fall, they would teach the single 3-hour credit course devoted to a research project involving the math majors and non-majors that had advanced.

There are a number of advantages of a Research Educator Math Liaison faculty member overseeing the research and educational directions of the math research experiences. Such a position is a more permanent position, thus permitting students to stay with the RE Math Liaison for the full year (or more) and establishing a continuity of research. The position would allow this faculty member to engage more meaningfully with students and it would enrich their teaching experience. The RE Math Liaison could dedicate more time and energy to developing and expanding creative curricula, while aiming to make their own research accessible to FRI students. The RE Math Liaison might hope to propose that the group work on "high-risk" problems in their field which would be risky to devote too much time to personally, but may lead to interesting results and questions through the students work.

MIC Stream, v2.0: Student Admission

In order to promote the courses to the students, we would hosts informational "Open Houses" and create a dedicated webpage on the FRI website. Students would then indicate their interest by using an online form that is part of our normal Stream Sort process where they select their top choices for their main FRI stream. A selection process would then need to be used if the demand exceeded the number of seats available. The criteria for selection would seek to evaluate a given student's enthusiasm, curiosity and aptitude for the math experience. This could be done by posing students with a problem, not unlike a recreational logic puzzle, that can be phrased without using mathematical vocabulary. The problem would pique the intellectual curiosity of students who may not realize that they had existing math interest and engage the students who do. A one-on-one interview with the instructor would give the students an opportunity to explain their thought process as they worked toward an answer to the puzzle or give a solution and give the instructor insight for evaluating the enthusiasm level of the students and their mathematics background.

MIC Stream, v2.0: Student Assessment

In order to track the impact of such a model, we assess the experience in a few different ways. Entrance and exit surveys would be administered to the students, after having obtained IRB (Institutional Review Board) clearance. The entrance survey would seek to ascertain their motivations for becoming involved and what their mathematics skill-set was prior to entry. The results of the exit survey would then help to determine what the students' perceived benefit of the course was and how their interest and confidence level in math had changed over the course of the experience. The FRI program as a whole already has an assessment strategy in place that uses multivariate analysis to determine odds ratios of academic outcomes for the students. Students within the MIC Stream would be a part of this analysis and would allow observations to be made about the potential impact of the experience. The students' activities after the math experience would be tracked so that involvement in other math experiences could be compared to students who did not do the program but had similar academic characteristics. Upper-division GPA could be monitored to observe any potential effects upon grades by this math experience. Lastly, the career trajectories of the students would be catalogued to see if there was an effect upon whether they went on to graduate and professional schools and what their ultimate career might be.

Conclusion

The FRI program offers a powerful model for improving the undergraduate research experience for students and addressing the challenges of STEM learning in the United States. We have shown the potential to extend these benefits more fully to mathematics students is possible through a small scale addition to the students' existing FRI undergraduate experience. We anticipate that scaling this approach up to incorporate more students will provide a meaningful experience in mathematics for them that applies to the natural and computer sciences. The cross-disciplinary nature of this approach is a high priority criterion to large granting agencies (e.g. NSF and HHMI) and smaller private foundations which seek to support efforts in developing mathematicians to think like scientists, and scientists to be mathematically literate so that they can contribute more effectively to the challenges and opportunities that are to come for society. Through this unique program and its potential adoption at other institutions, we hope to make a significant contribution to this same end.

Acknowledgments

This work was primarily supported by the Freshman Research Initiative at the University of Texas at Austin through the College of Natural Sciences, the Howard Hughes Medical Institute and the National Science Foundation. Gwendolyn M. Stovall was partially supported by both the National Science Foundation (Cooperative Agreement No. DBI-0939454) and the National Institute of Health (5R03HD068691). Any opinions, findings, and conclusions or recommendations expressed in this material are those of the author(s) and do not necessarily reflect the views of The University of Texas at Austin, the College of Natural Sciences, the Howard Hughes Medical Institute, National Science Foundation, or National Institute of Health.

References

[1] President's Council of Advisors on Science and Technology, Engage to Excel: Producing One Million More College Graduates with Degrees in Science, Technology, Engineering, and Math. PCAST STEM Undergraduate Education Working Group, S. G. J., J. Handelsman, G. Lepage, C. Mirkin, C.-c., eds. 2012.

[2] The Boyer Commission on Educating Undergraduates in the Research University, Reinventing Undergraduate Education: A Blueprint for America's Research Universities. http://naples.cc.sunysb.edu/Pres/boyer.nsf/ (accessed on December 2013).

[3] J. Graham, R. Caso, J. Rierson, J.-H.-H. Lee, The Impact of the Texas LSAMP Program on Under-represented Minority Students at Texas A & M University's College of Engineering: A Multi-Dimensional Longitudinal Study, **2** (2) (2002) F4B–1–F4B–6.

[4] Hobby Center for the Study of Texas at Rice University, http://hobbycenter.rice.edu/data (accessed on January 2014).

[5] 2012–2013 Statistical Handbook Quick Reference Guide, http://www.utexas.edu/academic/ima/sites/default/files/IMA_Pub_QuickReference_2012_Fall.pdf (accessed on January 2014).

[6] Rising Above the Gathering Storm Energizing and Employing America for a Brighter Economic Future, http://www.nap.edu/openbook.php?isbn=0309100 399 (accessed on January 2014).
[7] D. Lopatto, Undergraduate Research Experiences Support Science Career Decisions and Active Learning, *CBE Life Sci. Educ.*, **6** (4) (2007) 297–306.
[8] E.L. Dolan, D. Johnson, The Undergraduate-Postgraduate-Faculty Triad: Unique Functions and Tensions Associated with Undergraduate Research Experiences at Research Universities, *CBE Life Sci. Educ.* **9** (4) (2010) 543–553.
[9] S.H. Russell, M.P. Hancock, J. McCullough, The Pipeline. Benefits of Undergraduate Research Experiences, *Science* **316** (5824) (2007) 548–549.
[10] CNS College Fact Sheet, http://cns.utexas.edu/about/facts (accessed on January 2014).
[11] K. Procko, L.S. Sarah, Research as an Introductory Course: Engaging First-Year Students in Authentic Chemistry Research through the Freshman Research Initiative Program, in *Developing and Maintaining a Successful Undergraduate Research Program.*, Vol. 1156. Washington, DC: American Chemical Society, 2013 pp. 121–145.
[12] A. Collins, J.S. Brown, S.E. Newman, Cognitive Apprenticeship: Teaching the Crafts of Reading, Writing, and Mathematics, in *Knowing, Learning and Instruction. Essays in Honor of Robert Glaser*, Hillsdale, NJ, 1989, pp. 453–494.
[13] M.C. Linn, N.C. Burbules, The Practice of Constructivism in Science Education, (1993), 91–119.
[14] C.W. Bowen, A Quantitative Literature Review of Cooperative Learning Effects on High School and College Chemistry Achievement, **77** (2000) 116–119.
[15] W.J. McKeachie, McKeachie's Teaching Tips: Strategies, Research, and Theory for College and University Teachers, (1999), 158–166.
[16] D.E. Allen, H.B. White, Student Assisted Teaching: A Guide to Faculty–Student Teamwork, (2001) 134–139.
[17] J.L. Sarquis, J.L. Dixon, D.K. Gosser, J.A. Kampmeier, V. Roth, V.S. Strozak, P. Nelson, Student Assisted Teaching: A Guide to Faculty–Student Teamwork, (2001) 150–155.
[18] D.C. Lyon, J.J. Lagowski, A Study of the Effectiveness of Small Learning Groups in Large Lecture Classes, (2005).
[19] M.J. Chidister, F.H. Bell, K.M. Earnest, Student Assisted Teaching: A Guide to FacultyStudent Teamwork, (2001) 2–7.
[20] D.W. Johnson, R.T. Johnson K.A. Smith, *Active Learning: Cooperation in the College Classroom.* Interaction Book, 1988, pp. 9.3–9.18.
[21] R.I. Shear, S.L. Simmons, *Teaching Through Research: Five-Year Outcome Data from the Freshman Research Initiative at the University of Texas at Austin., 67th Southwest regional ACS meeting*, Austin, TX, 2011.
[22] H. Thiry, T.J. Weston, S.L. Laursen, A.B. Hunter, The Benefits of Multi-year Research Experiences: Differences in Novice and Experienced Students' Reported Gains from Undergraduate Research, *CBE Life Sci. Educ.*, **11** (3) (2012) 260–272.

[23] O.A. Adedokun, L.C. Parker, A. Childress, W. Burgess, R. Adams, C.R. Agnew, J. Leary, D. Knapp, C. Shields, S. Lelievre, D. Teegarden, Effect of Time on Perceived Gains from an Undergraduate Research Program. *CBE Life Sci. Educ.*, **13** (1) (2014) 139–148.

[24] M. Estrada-Hollenbeck, A. Woodcock, P.R. Hernandez, P.W. Schultz, Toward a Model of Social Influence that Explains Minority Student Integration into the Scientific Community, *J. Educ. Psychol.*, **103** (1), (2011) 206–222.

[25] M. Delbruck, in Knotting problems in biology, in Mathematical Problem in the Biological Science Proceedings of Symposia in Applied Mathematics. American Mathematical Society, 1962, pp. 55–63.

Chapter 11

Determining the Impact of REU Sites in the Mathematical Sciences

Jennifer Slimowitz Pearl

The National Science Foundation's Research Experiences for Undergraduates (REU) Sites and Supplements program in the mathematical sciences has provided funds for faculty to guide research opportunities for students for over 25 years. While the number of sites fluctuates slightly, there are currently approximately 55 active NSF REU sites in the mathematical sciences operating around the nation. The National Security Agency has also traditionally provided support for REU sites. Although there is published literature citing the benefits of undergraduate research in general, and although each REU site is required to conduct an individual project assessment, an overall notion of the impact of the cohort of mathematical sciences REU sites is not yet available. In the recent federal climate, which includes tighter budgets and more focus on evidence-based practices, the mathematical sciences REU program might be well-served by sites taking advantage of established tools in a coordinated way to assess its impact on participating students. This article attempts to (1) describe some of the prior work relevant to the mathematical sciences community that has been done to determine the impact of REU sites and of research for undergraduates more generally and to (2) describe a specific tool, the Undergraduate Research Student

MSC 2010: 01A80 (Primary); 01A67 (Secondary)

Self-Assessment, that might be useful going forward. While there are different types of evaluations that can be done (longitudinal, post and pre surveys, etc.), a careful examination of specific benefits of different types of evaluation is beyond the scope of this article. The aim here is only to illustrate the possibility of using already-developed tools to get a real sense of impact, without requiring a great deal of investment in time or money on the part of the organizer.

In recent years, there has been increasing emphasis on supporting training activities for students which are known to be effective [1–7]. Undergraduate research has been highlighted in the literature as a practice that improves students' probability of planning to enroll in a graduate program [6], increases retention [7], and increases students' skills, confidence, and motivation (various studies noted in [5]). Further, in [2], the authors have detailed best practices for implementing undergraduate research. While these references do not specifically address undergraduate research in the mathematical sciences, their results are meant to be applicable across fields and are enlightening.

Two publications from mathematical sciences professional societies address the impact of undergraduate research in mathematics. In 2006, the American Mathematical Society surveyed individuals who had participated in REU programs during the years 1997–2001 [4]. The survey, which received 262 responses, showed that all but one respondent found the REU to be either *Valuable* or *Somewhat Valuable.* A total of 82% of respondents entered graduate school, and of these 78% noted that the REU was a factor in their decision to do so. The survey also showed that for the majority of students, participation in an REU did not shorten their time in graduate school and did not influence their choice of thesis area. Also in 2006, the Mathematical Association of America published a report which lists costs and benefits of undergraduate research [3]. They cite costs to faculty and departments that relate to mentor time, and note "tremendous" benefits to students that include "independent learning, ... control over their education,... enriched... understanding of modern mathematics,... [and] presentation of results in written and oral formats."

Taking into account the literature cited above, a solid case has been made for the positive benefit of research for undergraduates in the mathematical sciences. That being said, to ensure the continuity, growth, and optimal development of the REU program, assessment needs to be enhanced so that outcomes are better understood. As per the current REU solicitation NSF 13-542 (http://www.nsf.gov/publications/pub_summ.jsp?ods_key=nsf13542), grantees are required to

> Describe the plan to measure qualitatively and quantitatively the success of the project in achieving its goals, particularly the degree to which students have learned and their perspectives on science, engineering, or education research related to these disciplines have been expanded.

Grantees handle this requirement in different ways. Some will survey students themselves, some hire outside evaluation experts to design and implement surveys, and some do neither. Many report on publications, presentations, and career paths of participants. Many have collected anecdotes that describe the real and compelling positive impact of the REU experience (see [4] for example). However, anecdotes are hard to quantify and hard to use in making programmatic decisions. It is also important to note that the different REU sites may have different specific objectives and different target populations, making it important to allow sites to customize their assessment to meet their needs.

There are some drawbacks to the current system whereby each mathematical sciences REU site develops an assessment plan that is independent of the assessment for other REU sites. For example, each site director must craft his or her own assessment plan or must pay an expert to do so, requiring an outlay of (at least) precious time and potentially of grant funds. Not all sites use survey instruments, and those which do have instruments that vary both in terms of their content and in how they are administered. In addition, given the heterogeneity in the generated data sets, there is no way to aggregate data across sites, and hence no way to develop a sense of the impact of the REU program as a whole.

There are resources that are available online that compile information about assessment of undergraduate research. For example, on the website http://serc.carleton.edu/NAGTWorkshops/undergraduate_research/assessment_pedagogy.html, Jill Singer from SUNY-Buffalo

State and Dave Mogk from Montana State University Bozeman describe several different assessment instruments and methods. One of these is the Undergraduate Research Student Self-Assessment (URSSA), developed by a team of researchers including Anne-Barrie Hunter, Tim Weston, Heather Thiry and Sandra Laursen at the University of Colorado. The URSSA is based on SALG (Student Assessment of their Learning Gains) which was originally designed for courses. SALG is also freely available. More about SALG can be found here http://www.salgsite.org/about. While URSSA is by no means the only tool available to assess undergraduate research, it will be highlighted in this article because it is already routinely used by principal investigators of NSF REU awards in some other disciplines, specifically the biosciences.

URSSA is based on the Student Assessment of their Learning Gains instrument and assesses how students feel they have benefited from undergraduate research experiences. It has been validated and tested and is available online http://www.colorado.edu/eer/research/undergradtools.html. Using URSSA, faculty members are able to customize a survey to meet their needs, administer it to students electronically, and retrieve the data for their own use. The website provides further links to a Frequently Asked Questions document and a link to the actual URSSA questions.

Using a freely available, and validated survey instrument such as URSSA could help REU site directors in several ways. It would save them the effort and cost of developing their own survey instrument. Accurate survey data would help them refine their programming to best foster student gains. And, if data could be aggregated across sites, this data would allow the mathematical sciences community and funding agencies to better understand the impact of the REU program as a whole. If there is a strong positive impact, this data can help make a case for the protection of the REU program in times of tight budgets or growth in better budget years.

URSSA is not customized for any discipline in particular, and most of the questions are appropriate for the mathematical sciences. The survey has questions in areas such as "Gains in thinking and working like a scientist: application of knowledge to research work" and "changes in attitudes or behaviors as a researcher." As an illustration, an excerpted set of questions is presented at the conclusion of this article. The survey does not measure how much students liked their REU experience, rather it aims to clarify how much they have gained from the experience. The URSSA developers have also instituted a "group" functionality which allows a set of REU directors to agree upon a set of common questions and to aggregate the responses to these questions across sites.

URSSA is being used by the REU Site programs funded by the Biological Sciences Directorate (BIO). A three-year pilot implementation has just been completed, and the BIO-funded REUs are now in a position to aggregate data from approximately 140 programs. The process of getting all programs to adopt a common assessment tool started approximately 4 years ago when the BIO REU principal investigators selected the URSSA tool over others. Pilot data shows positive gains from students, with approximately 800 students surveyed thus far.

During the summer of 2014, a group of approximately ten mathematical sciences REU principal investigators led by Kasso Okoudjou at the University of Maryland will use URSSA to survey their students. In addition, they will pilot the URSSA group functionality. It is hoped that this experiment — which requires neither a large expenditure of time or money — will begin to gather data that will illustrate impact of the individual mathematical sciences REU sites and the program as a whole.

Disclaimer: "Any opinion, finding, and conclusions or recommendations expressed in this material; are those of the author and do not necessarily reflect the views of the National Science Foundation."

Excerpt of questions from URSSA. Student responses are indicated on a Likert scale of the type none/a little/some/a fair amount/a great deal/not applicable. Not all questions will be applicable to all REU sites or to all students within a site. The survey is customizable so that REU site directors can add or delete questions as appropriate. There are a set of core questions which cannot be deleted in order to keep the instrument valid.

(http://www.colorado.edu/eer/research/documents/URSSA_MASTER_reviewCopy.pdf).

1. How much did you GAIN in the following areas as a result of your most recent research experience?

1.1 Analyzing data for patterns.
1.2 Figuring out the next step in a research project.
1.3 Problem-solving in general.
1.4 Formulating a research question that could be answered with data.
1.5 Identifying limitations of research methods and designs.
1.6 Understanding the theory and concepts guiding my research project.
1.7 Understanding the connections among scientific disciplines.
1.8 Understanding the relevance of research to my coursework.

2. How much did you GAIN in the following areas as a result of your most recent research experience?

2.1 Confidence in my ability to contribute to science.
2.2 Comfort in discussing scientific concepts with others.
2.3 Comfort in working collaboratively with others.
2.4 Confidence in my ability to do well in future science courses.
2.5 Ability to work independently.
2.6 Developing patience with the slow pace of research.
2.7 Understanding what everyday research work is like.
2.8 Taking greater care in conducting procedures in the lab or field.

3. How much did you GAIN in the following areas as a result of your most recent research experience?

3.1 Writing scientific reports or papers.
3.2 Making oral presentations.
3.3 Defending an argument when asked questions.
3.4 Explaining my project to people outside my field.
3.5 Preparing a scientific poster.
3.6 Keeping a detailed lab notebook.
3.7 Conducting observations in the lab or field.

3.8 Using statistics to analyze data.
3.9 Calibrating instruments needed for measurement.
3.10 Working with computers.
3.11 Understanding journal articles.
3.12 Conducting database or internet searches.
3.13 Managing my time.

4. During your research experience HOW MUCH did you:
4.1 Engage in real-world science research.
4.2 Feel like a scientist.
4.3 Think creatively about the project.
4.4 Try out new ideas or procedures on your own.
4.5 Feel responsible for the project.
4.6 Work extra hours because you were excited about the research.

References

[1] National Science and Technology Council. *Coordinating Federal Science, Technology, Engineering, and Mathematics (STEM) Education Investments: Progress Report.* Washington, DC: Executive Office of the President, 2012.

[2] J.E. Brownell, L.E. Swaner. *Five High-Impact Practices: Research on Learning, Outcomes, Completion, and Quality.* Washington, DC: Association of American Colleges and Universities, 2010.

[3] Committee on the Undergraduate Program in Mathematics. *Mathematics Research by Undergraduates: Costs and Benefits to Faculty and the Institution.* Washington, DC: The Mathematical Association of America, 2006.

[4] F. Connolly, J.A. Gallian. What Students Say About Their REU Experience. *Proceedings of the Conference on Promoting Undergraduate Research in Mathematics* (pp. 233–236). Providence, RI: American Mathematical Society, 2007.

[5] M. Crowe, D. Brakke. Assessing the Impact of Undergraduate-Research Experiences on Students: An Overview of Current Literature, *CUR Quarterly*, **28** (2008) 43–50.

[6] M. Eagan, S. Hurtado, M. Chang, G. Garcia, F. Herrera, J. Garibay. Making a Difference in Science Education: The Impact of Undergraduate Research Programs, *Am. Educ. Res. J.* **50** (2013) 683–713.

[7] President's Council of Advisors of Science and Technology. *Engage to Excel: Producing One Million Additional College Graduates with Degrees in Science, Technology, Engineering, and Mathematics.* Washington, DC: Executive Office of the President, 2012.

Chapter 12

The Mathematics REU, New Directions: A Conference at Mount Holyoke College, June, 2013

Giuliana Davidoff, Mark Peterson, Margaret Robinson and Yanir A. Rubinstein

Mount Holyoke College has hosted a summer REU in mathematics since 1988, and it seemed natural to us to mark these 25 years with some kind of celebration. Three of us have been active in the MHC REU over the years, while one of us participated as un undergraduate in the 1999 MHC REU. As one of the first mathematics REU programs, MHC's had stuck to what might be called "the early model," namely small teams of talented mathematics undergraduates confronting a problem of genuine mathematical interest with the close collaboration of a faculty mentor. As the organizers talked, though, they realized that in those 25 years the REU idea had not stood still.

The conference was organized to find out how the early model had evolved, and what those developments suggested for the future. The conference was funded by an NSF supplement grant DMS-1301735 and the Hutchcroft Fund of the Mathematics Department at MHC. There were about 60 participants in the conference from a wide spectrum of backgrounds and geographic locations, including

MSC 2010: 00B25, 01-06 (Primary)

current and former REU site directors and NSF officiers, university administrators, current and prospective REU faculty mentors, graduate and undergraduate students. The two days of the conference took two different thematic directions. The first day was "New Formats for the REU," that is to say, new ways to organize research-based education. The second day was "REUs and the Culture of Mathematics," observations on how the acceptance of research-based education might have changed American mathematics itself. Because changes in the culture of mathematics have, to some extent, been reflected in new models of research-based education, a clean separation of these two themes is not possible. In fact, arguably the most exciting developments of the conference came at the place where these themes intersect.

Each day was organized into panels, in which panelists made short presentations, followed by discussion. Discussion continued over lunch and dinner, and a final session on the second day attempted to capture ideas that had arisen informally outside the more formal proceedings. There was also one longer presentation each day, the first by Ravi Vakil (Stanford), the second by Carlos Castillo-Chavez (Arizona).

The panelists and their presentations are listed below in an overview of the conference.

Panel I: New Formats for REUs

I.1. Daniel Cristofaro-Gardner (UC Berkeley): The UC Berkeley Summer Research Program in Mathematics for Undergraduates
I.2. Tamas Forgacs (Fresno): FURST — transcending the time-space limitations of a traditional REU
I.3. Haynes Miller (MIT): A project laboratory in mathematics
I.4. Ravi Vakil (Stanford): Experimenting with new ideas for REUs
I.5. Josh Beckham (UT Austin): One Institution's Approach: How the University of Texas at Austin Merges Research and Teaching through the Freshman Research Initiative
I.6. Frank Coale (Maryland): The Gemstone Program: Maximizing Learning Through Team-based Interdisciplinary Research

I.7. William Yslas Vlez (University of Arizona): REUs to retain mathematics majors

I.8. Jennifer Slimowitz Pearl (NSF): Taking a broad view: multiple models of REUs in the math community and what we might learn from others

I.A. Presentation: Ravi Vakil (Stanford): Planning REUs with intentionality: Ideas from elsewhere in mathematics

Panel II: REUs and the Culture of Mathematics and Statistics

II.1. Carlos Castillo-Chavez (Arizona): The power of student-driven REU programs: an alternative model of learning through research

II.2. Donal O'Shea (New College): Types of research suitable for REUs

II.3. Yanir Rubinstein (Maryland): Underrepresented areas in REUs

II.4. Ravi Vakil (Stanford): The spirit of collaboration and free exchange of ideas in the mathematics community, seen through REUs, and other topics

II.A. Presentation: Carlos Castillo-Chavez (Arizona): Pathogens, Disease Evolution and Mathematics: Challenges and Opportunities

II.B. Open discussion

In the following report the speakers will be referenced by the numbering scheme above, but the point at issue will also be found more fully presented, in most cases, in their written contributions to this volume.

Themes

A number of themes emerged during the conference that cut across the divisions foreseen by the organizers. This section seeks to identify them and to note interrelations among them. These interrelations preclude a neat division into discrete categories.

General Success of the REU Concept

The NSF's REU program was successful right from the start, as evidenced by its quick popularity and increasing visibility over the years and decades. A recent study, the 2013 DMS Committee of Visitors' Report, said "The REU program is a highlight of the [DMS] workforce program ... and is an exemplary program in its broader impacts." (Pearl, I.8.) [1, p. 16]. Take the MHC program as representative: almost 300 students have participated in the MHC REU over 25 years, and of these, many have gone on to be professional mathematicians or other mathematical scientists. That translates to thousands of students in the mathematics REU program nationally to date, and hundreds of professional mathematicians.

Normalcy of Undergraduate Research in Mathematics

Early accounts (some essays reprinted in Senechal, 1991, for instance [4]) recalled a time when college mathematics departments seemed to be largely service departments for the other sciences, where the undergraduate students who persisted beyond the introductory service courses were "survivors," the few who would ever do research and finally learn what mathematics is. This is a notion that is startling to read today, almost incomprehensible in the context of mathematics REU institutions. At such institutions all science students are familiar with the notion that mathematicians do research. That seems to be a success of the REU program that has gone almost unremarked — a large population of undergraduate students now understands better what mathematics really is.

REUs and Networking

Experience has revealed certain recurring problems with the original REU models. Tamas Forgacs (Forgacs, I.2) described a pilot program,

Faculty and Undergraduate Student Research Teams (FURST), that builds on the initial model of the REU and seeks to address some of its space/time shortcomings. Faculty–undergraduate research teams would be identified at home institutions *before* the summer. Undergraduates would go to the REU having been already prepared at their home institution for the summer's work. Faculty-undergraduate teams would continue to function after the summer in putting the summer's work into good shape, writing a paper, presenting at a conference, etc. In this way faculty at PUIs (primarily undergraduate institutions) without REUs, and perhaps without much encouragement to do research, would enjoy the stimulation of collaboration at a distance. Furthermore, the time constraints at both ends of the summer, getting a problem off to a good start and then finishing it up, would be more manageable. Undergraduates would benefit from a more realistic timetable and multiple mentors. This coordination of research at PUIs could be as beneficial to faculty as to students. The REU structure provides a niche for talented leaders of such programs that can be very satisfying professionally, perhaps even more so in this extended form.

REUs at PUIs and R1s

Initially REUs at primarily undergraduate institutions (PUIs) outnumbered REUs at first-tier research universities (R1s) by roughly 3 to 1 (O'Shea, II.2). It is understandable that, relatively speaking, the REU idea should have looked most interesting to PUIs. As the idea caught on, however, R1 institutions became a more important constituency, and began developing their own models, often not funded specifically by REU money but by other more or less equivalent mechanisms, or in-house. In the examples we heard about, it was in large part undergraduate demand that led to the summer programs.

The availability of graduate students at R1 institutions changes the REU picture quite substantially. Several of our panels dealt with this development (Gardiner, I.1; Vakil, I.4; Rubinstein, II.3). The graduate students in these programs had almost complete independence in how they conducted them. They were the ones choosing the problems, choosing the undergraduates, and working with them. It was noted with gratitude (and perhaps unease?) that graduate students were willing to direct undergraduates in summer programs for very little money. The problems undertaken in these programs were not in fields typical of REUs at PUIs: rather they sometimes reflected the topics of the graduate students' own research, making direct contact with the R1 research culture. In effect, it was a way of breaking out of the usual undergraduate curriculum. Faculty advised the graduate students as necessary, gave colloquium talks to the research groups, and engaged the undergraduate students in social settings.

The growing expectation that undergraduates should work on research problems will potentially affect the academic job market. Graduate students who have already had the experience of guiding undergraduates, or who have had a strong REU experience themselves as undergraduates, might look especially attractive to PUIs or even R1s hiring new faculty. The Mount Holyoke Conference included attendees who were new faculty expecting to direct their own REUs in the immediate future. For instance, Alexei Oblomkov (UMass Amherst) who attended the conference, applied for an NSF CAREER grant shortly after the conference, modeling the educational component of his proposal based on ideas raised in the conference. His proposal was funded (NSF DMS-1352398).

The choice of young faculty to direct undergraduate research, possibly at the expense of their own research, could be a questionable career strategy. Those involved in the tenure process, from mathematical colleagues, to colleagues in other departments, to administrators, should know the arguments for why this is a good thing to do.

Intention of REUs: Faculty

Ravi Vakil (Vakil, I.A) raised questions of what REUs are for, in a presentation that was also a group discussion. What is in it for the faculty?

- The reward of seeing student success — like having children!
- Investment in the future

- Within the institution, facilitates connections to other sciences (we do research too!)
- Opens horizons, new possibilities, can try a new area
- Can lead to connections with other institutions, breath of fresh air
- Culture of excitement, more activity
- Holistic: teaching/research/service all happening together
- Get past constraints of undergraduate curriculum
- (For graduate schools): recruit incoming graduate students
- (For graduate schools): more things for graduate students and faculty to talk about
- (For graduate schools): graduate students see better what the mathematical life is like, not just mathematics
- Social, not individual

The value of these things should be built into the system and its reward structure: tenure, salary, etc.

Intention of REUs: Students

Ravi Vakil (Vakil, I.A) also asked what was in the REU for students, distinguishing what students might want, and what faculty might want for them. (Some students attending from the current MHC REU suggested that these categories were not necessarily distinct.) The following things were noted, first what students perhaps think they want:

- Better CV
- Money (in lieu of a job, probably)
- REU students make better graduate students
- Relatively risk-free way to try out the mathematical life
- Meet other mathematics-oriented students
- Go beyond undergraduate curriculum
- Build confidence (N.B. There is also the risk of losing confidence!)

And then, things that students might not anticipate but that are among the conscious intentions of their mentors:

- Positive experience of engaging mathematics, doing mathematics
- Community: build relationships among mentors and peers

- Communication: work on presentation skills, speaking and writing
- Think about graduate school
- Encounter peers with more experience
- Build self-knowledge about abilities
- Foster a diverse mathematics community
- Encounter diverse ways of learning
- Make a sustained effort: hit a roadblock and work to get past it
- Integrate newcomers into the mathematical life

Effect of REUs on the Mathematics Profession

The increasing role of research experience in undergraduate mathematics education suggests that we ask what the effect has been on the mathematics profession. Remarkably, there seem to have been no formal studies of this (Pearl, I.8). Yanir Rubinstein suggested (Rubinstein, II.3) that among REUs there are overrepresented fields, like number theory, graph theory, and combinatorics, and underrepresented fields like geometry and analysis. He presented statistics to illustrate this point. No doubt it is easier to get started in some fields than in others, doing computer experimentation, for example. Some fields require more preparation even to comprehend their problems. But it is worth noticing that REUs might introduce a certain imbalance into American mathematics.

In a separate observation he suggested that REUs tend to foster a spirit of generosity and cooperation among the participants, and this spirit might thus characterize some fields of mathematics more than others, because of the uneven representation of fields in REUs.

Donal O'Shea (II.2), noted that younger mathematics professors now have experience with REUs, and that some fields, combinatorics and graph theory, for example, have been greatly stimulated. He also saw a trend toward more collaborative mathematics, a trend exemplified by REUs, contradicting the "lonely genius" model. Other roughly contemporaneous mathematical movements and institutions, MSRI for instance, indicate the increasingly collaborative nature of mathematics. If this kind of collaboration among mathematicians could extend also to community colleges it would be good for society.

Most college students still attend institutions where they never even see what a mathematician might call real mathematics.

Other REU Models

Jennifer Slimowitz Pearl (Pearl, I.8) described the NSF evaluation process for REU proposals, and mentioned some recent innovative ones that have been funded. Illinois State involves pre-service and in-service teachers in a mini REU for local high school students, aiming to help teachers become teacher-scholars. North Carolina State does modeling and industrial applied mathematics, in partnership with companies around the country, using Skype to stay in touch. (Worcester Polytechnic Institute, University of Maryland, Baltimore County and Harvey Mudd College also partner with companies to find applied problems.)

Jennifer Slimowitz Pearl cited developments in other fields that might be relevant to mathematics REUs. Biologists have developed a common assessment tool (see www.bioreu.org), and "just in time review" by graduate students of undergraduates' posters and papers. The assessment tool produces aggregate data for REUs in biology as a whole, across the country. Geosciences have developed models for using communications tools to bring students and faculty together online, rather than bringing everyone to the same location. See Hubenthal and Judge, EOS (2013), "Taking Research Experiences Online." [2].

REU mini-grants are given by the Center for Undergraduate Research in Mathematics (CURM) administered by BYU. The NREUP program of the MAA supports the participation of mathematics undergraduates from underrepresented groups in focused and challenging research experiences to increase their interest in advanced degrees and careers in mathematics.

Melissa Zhang, a student at the current MHC REU, suggested existing REUs might allow affiliations with students not directly participating, possibly high school students, attending through occasional Skype sessions. These students would experience some of what was happening, in almost an observer status, a much lower level of commitment, but perhaps good for them if they are uncertain about their degree of interest. It would be a "preliminary REU."

Migration of the REU Idea to the Curriculum

Several panels (I.3, I.5, I.6) described courses or whole curricula that were essentially REU-like.

Haynes Miller (I.3) described an MIT course in which nine teams of three undergraduates work on their choice (from a long list) of three open-ended problems in mathematics for 4 weeks each. The experience is REU-like, and students are free to consult other sources except what students have done with these problems in the past. The course puts a strong emphasis on oral and written exposition. There is not room to accommodate students from outside the mathematics department, and the course is strongly recommended to mathematics majors. The course has appeared in MITOpenCourseWare.

Josh Beckham (I.5) and Frank Coale (I.6) described large-scale interdisciplinary curricula at Texas and Maryland respectively, involving hundreds of undergraduates in a research-based approach to a college education. The Texas program, Freshman Research Initiative, is a 3-semester program. The Maryland program, Gemstone, comprises all 4 years. In each of them teams of undergraduates ("research streams" of 15–35 students at Texas, close teams of 3 or 4 at Maryland) become involved in research problems which they, in part at least, can define. They have access to mentors throughout, and finish with a polished exposition of results. Gemstone evolved in response to corporations' view that their new hires lacked the skills of teamwork, and accordingly teamwork is a conscious aim of the program, facilitated by trained professionals. Graduates of these programs seem to have benefited greatly, as evidenced by various statistics.

The President's Council of Advisors on Science and Technology has written a report, "Engage to Excel," strongly advocating the kind of research-based curriculum (they call it "discovery-based") considered above (Pearl, I.8) [3]. Appendix I of that document is a bibliography of recent studies of such programs, including REUs.

The Texas and Maryland programs are interdisciplinary within the sciences, including mathematics in principle. In fact, however, the students typically define problems that do not involve mathematics in any deep way, and perhaps hardly involve mathematics at all. This seems to be an interesting phenomenon, perhaps an opportunity for mathematicians to find some opening. The programs are thriving and successful, and there can be little question that many problems of interest to students could benefit from mathematical modeling of some kind, even if they do not think of it themselves. And conversely, mathematics itself can benefit from confronting applied problems. It appears that the REU idea, but not yet the mathematics REU idea, has made the transition to the curriculum in this case.

Migration of the REU Idea to Student-Chosen Problems

An intriguing development was described by Carlos Castillo-Chavez (II.1, II.A) and seconded by William Velez (I.7), namely the use of REUs to recruit and foster student research on topics of their own choosing or problems otherwise particularly interesting to them. These will typically be applied, interdisciplinary problems. Both Castillo-Chavez and Velez have used this format to draw in categories of students, especially but not exclusively Hispanics, who might not have been so interested in mathematics without the clear relevance to their lives that their approach provides.

Castillo-Chavez and Velez seem to be exploiting the opportunity noticed in the previous subsection, the opportunity to involve students in mathematics through specifically applied mathematics.

Exposition, Oral and Written

Donal O'Shea (II.2) pointed out that the earliest discussions of the potential for undergraduate research in mathematics, predating the REU program, usually put little emphasis on the importance of exposition, focusing largely on the mathematics itself, perhaps because only new results would even require exposition, and the aim of the program was to produce new results. The history of the REU program has been one of increasing emphasis on exposition and presentation, and as a result (and to some extent also a cause) there are now many venues available for undergraduates both to speak (MathFest, MAA meetings) and to publish mathematics (Involve, Rose-Hulman Undergraduate Math Journal, and Ball State Mathematics Exchange were mentioned).

All participants in REUs should expect to improve their presentation and writing skills. This would be a good outcome no matter the success or failure of the project itself. The practical value of being a good speaker and a good writer on technical subjects is obvious.

Ravi Vakil suggested that a good exposition of some topic could even be the main goal of an REU, putting the emphasis on exposition right from the first day. (It is worth noting Edward Packel's contribution "On the Value of 'Expository Research' for Mathematics Undergraduates" [4, pp. 73–76] in L. Senechal, 1991, suggesting essentially the same thing.)

Interdisciplinarity

Early mathematics REU programs seem to have been more concerned with mathematical purity than many programs are today. Velez (Velez I.7) described urging his mathematics students into research situations that aren't purely mathematics but that use mathematics. This, he suggests, will keep them interested in mathematics, by making it useful to them. Double major, he tells them: do mathematics and something else. He emphasized the importance of learning skills of scientific computation for use in some other discipline. Castillo-Chavez (Castillo-Chavez II.1) has also emphasized interdisciplinary problems with mathematical content, problems that arise out of students' own interests, even if these initially, perhaps, have not so much to do with mathematics.

On the other side, interdisciplinary programs in other sciences, like those at Maryland and Texas (Beckham I.5, Coale I.6), have perhaps unrecognized mathematical content without the participation of mathematicians to see the possibilities. This seems like a missed opportunity, since mathematics is often the most efficient way to import the insights of one discipline into another.

Ravi Vakil (Vakil, II.4) spoke strongly in favor of the importance of every kind of mathematics, from the purest to the most applied, in undergraduate research situations. Administrators will need metrics for this. There are many different meanings for "doing research."

Experimentation

It was remarked that other scientists can put students in labs and start them doing something, but we don't have labs. On the other hand, there are areas where a kind of experimentation is possible, generating examples, for instance. Computers make possible new kinds of mathematical experimentation. It was noted that mathematics may have become more experimental in the last generation, partly because of REUs, perhaps, and partly because of technology.

Diversity

A marked change from the early years is a new understanding that research experience is not just for mathematically elite students, and that it may have special importance for student groups heretofore underrepresented in mathematics. This theme came up repeatedly (Castillo-Chavez, II.1; Velez, I.7; Pearl I.8; Beckham, I.5; Coale, I.7). If there is a movement, though, it is in its early stages. Castillo-Chavez might have seen more Hispanic students in his program than there have been in all other programs nationwide combined. The research-oriented curricula at Maryland and Texas are diverse in the same proportions as the student population they draw from, but they are not particularly mathematical in their emphases. It seems clear that the life experience of many bright students would predispose them to be interested primarily in applied problems, and that pure mathematics would seem to all but a very few of them to be completely unmotivated. Only over time would a culture of more general and abstract interest in mathematics take root in such communities, but this could be an important development in the long term. The leaders in these developments were vocal in worrying about their continuation to the next generation.

The tensions within the mathematical community: teaching/ research, say, or pure/applied, are present within the REU program because all aspects of mathematics are found there in miniature. The conference seemed to encourage a respect and tolerance for all sides of such issues, perhaps because the participants had seen all sides in their own REU experiences. This kind of toleration and respect may be yet another way that the REU program will contribute to the mathematics discipline, a contribution to the climate in which mathematics is done. The tolerance and respect that comes from bringing people together to do mathematics, while they also cooperate in other ways, should extend to other kinds of diversity issues, such as sexism, classism, etc. Anecdotes suggested that REUs have also been useful in this way.

References

[1] COV, Report of the 2013 Committee of Visitors, Division of Mathematical Sciences, National Science Foundation, 2013. http://www.nsf.gov/mps/advisory/covdocs/2013_DMS_COV_Report.pdf.

[2] M. Hubenthal and J. Judge, Taking NSF's Research Experiences for Undergraduate (REU) Sites Online, *Eos Trans. AGU*, **94** (17) (2013) 157–158.

[3] PCAST, Engage to Excel: Producing One Million Additional College Graduates with Degrees in Science, Technology, Engineering, and Mathematics, 2012, http://www.whitehouse.gov/sites/default/files/microsites/ostp/pcast-engage-to-excel-final_feb.pdf.

[4] L. Senechal (ed.), *Models for Undergraduate Research in Mathematics.* MAA Notes Series No. 18, Mathematical Association of America, 1990.

www.ingramcontent.com/pod-product-compliance
Lightning Source LLC
Chambersburg PA
CBHW050638190326
41458CB00008B/2326

* 9 7 8 9 8 1 4 6 3 0 3 1 3 *